D1363705

sheep's
miscellany

a collection
of
truths and
trivia

Karen Gee

PIER
9

contents

Conclusion

References

introduction

'Like being savaged by a dead sheep.'

These words were uttered by British politician Dennis Healey after he'd been verbally attacked by fellow politician Geoffrey Howe. As Howe no doubt did, we all readily identify this as a cool yet rather scathing put-down. Equally, if someone were to accuse you and your friends of following the latest trends 'like a flock of sheep' it's likely you'd take this as an insult, as indeed it would be intended. And many a New Zealander has had to endure the endless rounds of derogatory and insinuating sheep jokes hurled at them on account of their country's famously large ovine population.

In the popular imagination sheep don't exactly have a reputation as being robust or resilient. To most of us they're the cute, cuddly and harmless creatures that form an integral part of our Western childhood, through nursery rhymes and storybooks. But not too many of us look beyond

this fluffy veneer. As you'll read within the pages of this book, sheep have been an important part of our lives for thousands of years, giving us meat, milk, wool and other by-products. And, as outlined in the second chapter, they played a seminal role in the development of human society.

You'll also discover that sheep are not the brainless beasts we've come to see them as. These woolly ruminants are, it seems, more intelligent than is commonly believed – they are able to recognise the faces of other animals within their flock, they've been known to show compassion, and have displayed great ingenuity in problem solving. (Granted, they are yet to be seen in university lecture theatres taking Engineering 101, but don't let that stop them trying to find a unique way of getting to the greener grass on the other side.)

I hope you enjoy this stroll through the pasture with what is one of history's more misunderstood animals, and through dipping into these pages I hope you learn to more fully appreciate all things ovine.

Woolly facts

Altogether too many sheep.
– George Bernard Shaw, on visiting
New Zealand in 1934.

A quick language lesson

Ewe: a female sheep

Ram: a male sheep (called a 'tupp' in some countries)

Wether: a castrated male sheep

Hogget: a one-year-old sheep

Two-tooth: an 18-month-old sheep who has — you guessed it — two teeth

Hermit: a sheep that has never been mustered

Gummy: an older sheep that has lost all its teeth (around 8–10 years old)

Mismother: a lamb that has lost its mother

Tallow bag: an overly fat ewe

Sharkmouth: a sheep whose top jaw protrudes over the bottom jaw

Pigmouth: opposite to a sharkmouth — the bottom jaw protrudes beyond the top jaw

Doubledecker: a sheep that has missed a year's shearing and has two years' wool on its back (identified by a distinct mark in the fleece, where one year's growth finishes and the next year's begins)

Mulesing: a process whereby folds of skin are removed to prevent maggots of blowflies infesting a sheep's skin with what is known as either sheep strike or flystrike

A day in the life

Pregnancy: Varies with breed. Lambs can become pregnant before they are one year old. Annual lambing is common, though it is possible for ewes to give birth every six to eight months.

Gestation: Approximately 145 days.

Number of lambs per pregnancy: Can be between one to three lambs, with twins being common.

Life expectancy: Increases with the size of the sheep, but the average is around 10 to 12 years.

Chromosome number: 54

Years of productivity: Again, this varies. A ewe will be most productive between three and six years of age, though can be productive past 10 years if well fed and cared for.

Average temperature: 39.1°C (104.0°F).

Average respiration rate: 16.

Average pulse rate: 80.

Daily consumption levels : food = 0.9-2.04 kg (2-4.5 lb) water = 0.5-1.6 litres.

Daily waste volume: fecal = 0.9-2.95 kg (2-6.5 lb) urine = 10-40 ml per kg body weight.

How many?

There are over 1 billion sheep currently roaming the planet. Of those, approximately 110 million reside in Australia, while around 40 million call New Zealand home. The United Kingdom plays host to 36 million sheeply souls, and China has around 158 million ovine friends. The United States, a minor player in the sheep stakes, has only 6 million woolly residents. As of 2002, sheep in developing countries accounted for approximately 64 per cent of the world's ovine population.

What kinds?

There are around 900 different breeds of domestic sheep throughout the world. Breeds are differentiated by many factors: their fibre (hair; or fine, medium or coarse wool), what they're raised for (meat, wool and milk), the colour of their faces, and other physical characteristics. More specifically, domestic breeds are divided into categories based on the conditions and uses for which they have been bred and adapted. These categories include: fine wool breeds;

mutton-type breeds; short-tailed breeds — renowned for their high rates of reproduction and often used for crossbreeding in order to increase the number of lambs born per ewe in other breeds; fat-tailed breeds, which are primarily kept for their milk-producing abilities; and haired breeds, found mainly in tropical areas and mostly bred for their meat.

On the wild side, the majestic Bighorn sheep, native to North America, once grazed the Rocky Mountains from southern Canada to Colorado. These days the breed exists only in small bands, in areas protected by the relative inaccessibility of its habitat. They can be found on grassy mountainous slopes, in alpine meadows, or amongst the foothills of rocky cliffs. Recognizable by the huge, curled horns of the ram, Bighorn sheep are one of only a handful of wild breeds remaining in the world, the others being: Dall's sheep, found in northern Canada and Alaska; the Mouflon from Central Asia; the Asian Mouflon, found in western Asia; the Urial, of Afghanistan and Pakistan; and the Argali, of eastern Asia.

For the Bighorn male, size does matter: horn size, along with age, plays an important role in establishing male dominance. In fact, horns can weigh as much as 13 kilograms (28 lb) and reach up to 83 centimetres (33 ½ in) in length. Bighorn rams famously fight it out over the right to mate with a particular female. The loud, reverberating clanging noise made by their clashing horns can be heard up to several kilometres away. Now sadly quite rare, the Bighorn was listed as an endangered species in 1998 by the US Fish and Wildlife Service.

One of the earliest ovine arrivals to Britain was the Soay. The most primitive breed in Europe, hailing from the remote island of Soay in the St Kilda archipelago off the Scottish coast, the breed is considered a valuable research subject for scientists, who have been studying it since the 1950s. What makes these sheep on their quiet, windswept isle so special? Their extreme isolation, lack of predators and the absence of human interference or influence mean Soays offer a rare link to the first domesticated sheep.

Champions of the Cotswold sheep are so taken with the breed they call it 'the Lion of the Cotswolds'. During medieval times, fleece of the Cotswold sheep was exported widely, mostly to Lombardy and Flanders, where it gave rise to the lyrics of a 12th century Flemish song: 'The best wool in Europe is Cotswold and the best wool in England is Cotswold'. (Clearly, lyricists of the day were not as inspired by the breed and its wool as the weavers were.) The breed's long, naturally wavy fleece contributed to England's wealth during the Middle Ages, and in particular to that of the Cotswold region where 'wool churches' and houses still stand as testament to the breed's pivotal role in the area's economy.

 But where's Australia's renowned Merino in all of this? Don't worry — simply turn to the third chapter.

Greedy guts

If you're looking for a great lamb recipe for your next family gathering, you won't find it here, as it's not our stomachs that are the subject, but the sheep's. Sheep are classified as ruminants, which means they chew cud, and they have four stomachs (or, more precisely, four compartments to their stomach). And if no one's ever explained to you exactly what cud is, then here's a quick introduction to this somewhat dubious goo: it's bolus, a substance consisting of already-consumed and swallowed food that's been regurgitated.

So why do sheep ruminate? It aids their digestion. Or maybe they're simply whiling away the hours, thinking of ways to free themselves of the monotony of the pasture. After all, they're more intelligent than you might think – just take a look at the fourth chapter of this book.

Biggest, smallest, weirdest

The world's largest sheep station, Commonwealth Hill, is located in the South Australian outback, not far from the opal mining town of Coober Pedy. It ranges over a vast 10,567 square kilometres (4080 square miles), or roughly seven times the size of the greater London area. Current sheep stock includes 55,000 Merino rams and 22,000 ewes.

The world's biggest sheep is so large you can take a look at it from the inside. At over 15 metres (49 ft) tall and 18 metres (59 ft) in length, and weighing in at a hefty 98 tonnes, the Big Merino towers over the New South Wales rural town of Goulburn. Constructed of a concrete-covered steel frame, the giant ruminant officially opened as a major tourist site in 1985, as a nod to the surrounding region's wool industry past. But the Big Merino doesn't depict just any old sheep – it was apparently modelled on local hero Rambo, a prize-winning stud ram from nearby Bullamallita Stud.

Once you've climbed the stairs into the sheep's innards, you can wander through the three floors, which house a display recounting the history of the Australian wool industry. You can even peer out through the ram's eyes for an elevated view of the local area and then make your way to the souvenir shop – conveniently located near the exit (under the tail, that is).

The largest breed of sheep in the world is the Lincoln, sporting what some consider to be the most luxurious fleece.

Their heavy, twisted locks – giving the creature a laid-back Rastafarian appearance – can measure as long as 38 centimetres (15 in). Originating on the east coast of England, the average weight for a mature Lincoln ram falls between 113 and 160 kilograms (249–353 lb), while ewes average between 90 and 113 kilograms (198–249 lb).

One of the most, well, unique-looking sheep is the Manx Loaghtan. Native to the Isle of Man, the breed's name is derived from the Manx words to describe the burnt appearance of its fleece – lugh (mouse) and dhoan (brown). This very rare, primitive and somewhat unfortunate-looking mountain sheep commonly has four horns, though it has been known to have six.

If you like 'em small, then take a trip to the Shetland Islands, off the north coast of Scotland. There, not only will you find the famously tiny Shetland pony, you'll also come across one of the world's smallest breeds of sheep, the inventively named Shetland. But small doesn't mean boring: the Shetland comes in one of the widest colour ranges of any breed. Also vying for the 'world's smallest' badge is the American Miniature, with mature ewes measuring roughly shoulder to shoulder with an adult beagle at around 40–46 centimetres (16–18 in), and the breed as a whole weighing in at a dainty 22.5 kilograms (50 lb) on average – a similar weight to an adult labrador.

Did you know?

Bighorn rams can go all night ... and all day: they have been observed locked in head-to-head combat over a female for more than 24 hours. Perhaps a nice bouquet and box of chocolates might have done the trick instead?

Global Sheep & Lamb stocks - 2005 (million head)

People's Republic of China	170.8	New Zealand	40
Australia	106	United Kingdom	35.2
European Union (15 nations)	98.7	South Africa	25.3
		Turkey	25.2
India	62.5	Pakistan	24.9
former Soviet Union	64.0	Nigeria	23
Iran	54	Spain	22.5
Sudan	48	All Other	337.3
		Total	**1,079.7**

Top Five's

Top Five exporters of Sheep skin with Wool
(metric ton)
1   Australia   81,440
2   United Kingdom   54,151
3   Spain   15,721
4   France   15,942
5   Turkey   9,537

Top 5 exporters of sheep by quantity (head)
1   Australia   3,397,140
2   Syrian Arab Republic 2,132,939
3   Romania   2,008,818
4   Sudan   1,087,124
5   Hungary   751,828

Top 5 importers of sheep by quantity (head)
1   Saudi Arabia 2,379,159
2   Italy 1,715,426
3   Kuwait 1,481,290
4   Greece 951,476
5   France 438,514

a stroll through history's pastures

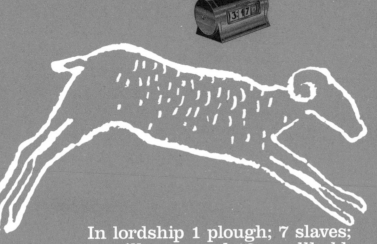

In lordship 1 plough; 7 slaves;
12 villagers and 12 smallholders
with 9 ploughs. Meadow, 16
acres; pasture 2 leagues long
and 1½ leagues wide; woodland,
8 furlongs. 2 cattle; 40 sheep

– Domesday Book,
entry for Hemyock Castle, 1086

Taming the wild beast

Today's domestic sheep *(Ovis aries)* is believed to be descended from the Mouflon *(Ovis orientalis; O. musimon)* of Central Asia. Sheep were one of the first animals to be domesticated by humans, around 10,000 years ago. Archaeological evidence of the earliest domestication of sheep has been found in the region of the Fertile Crescent (a historical Middle Eastern region extending from the Mediterranean Sea through to the Persian Gulf), in areas including northern Iraq and southwestern Iran, and at Jericho, believed to be the oldest continually inhabited settlement in the world.

While you might not think this taming of sheep has much bearing on our lives today, the domestication of livestock such as sheep was to pave the way for the development of more complex societies – in fact, if our ancestors hadn't corralled our hoofed friends into pens, not only might the paper on which this book is printed never have been invented, but you probably wouldn't have the leisure time in which to read.

Animal domestication helped make possible the spread of civilization: prior to this happening, people led a largely hunter–gatherer lifestyle, so the majority of everyone's time was taken up with procuring food. Once animals were domesticated, a more plentiful and protein-rich food source became more easily available. And since this wonderful food source was concentrated into a few pens, a greater amount of food could be generated by fewer people. This freed up some of the population, allowing them both to develop specialist skills and to devote time to contemplating their navels and thus creating and inventing. Theoretically, we could therefore thank the ancestors of Mary's little lamb for the invention of the wheel, Shakespeare's soaring verse, and Cliff Richard's memorable lyrics 'But I ain't losing sleep and I ain't counting sheep' (on second thought, perhaps Cliff should shoulder the blame for that one himself).

It is not known how domestic livestock arrived in Britain, though archaeological clues from the upper Thames area reveal that both farming and hunter–gatherer lifestyles existed simultaneously in the period 6300–5500 BC. Evidence has been uncovered to show domesticated sheep were being raised during the Early Neolithic Age (c. 4000–3000 BC), starting the Brits on a path that would eventually lead to what some consider their greatest culinary gift to the world – the lamb roast. The Domesday Book, the main volume of which is written on sheepskin parchment, records that by 1086 there were more sheep in England than all the other livestock added together.

. . . .

Did you know?

In 11th century England, countryside considered pasture was recorded in the Domesday Book in a number of different ways; in Essex, for example, land size was calculated according to how many sheep the land in question could support.

. . . .

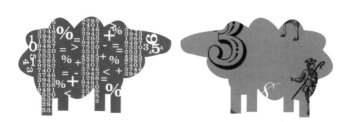

Making their move

While the mighty Merino has dominated the Australian wool industry, the breed's origins are linked with the spread of sheep and wool across the New World. It is generally accepted that the highly-prized Merino was developed in Spain during the eighth to 13th centuries and helped build that country's substantial wealth and power at that time. The national Council of the Mesta, a powerful collection of sheep owners in medieval Castile, maintained a monopoly over the breed, and exportation of sheep was forbidden. Wool was traded through ports closely monitored by a local shipping authority (another powerful association), and its export to places such as England and Flanders became a valuable source of income for Spain.

So lucrative did wool become for the Spanish that during the 15th century Queen Isabella and King Ferdinand used income from the industry to finance the exploratory voyages of conquistadors such as Columbus and Cortés. When Cortés began exploring Mexico and what would become the southwestern region of the United States, he took sheep with him – these sheep are believed to be the ancestors of the Navajo Churro breed, the oldest surviving breed in the United States.

The first sheep to make it to Australia arrived with the First Fleet in 1788, and the first Merinos – a total of only 13, purchased from the Spanish – set foot on Antipodean soil in 1797.

Starry, starry sheep

In Western astrology, the ram is the symbol of the first sign in the zodiac, Aries. If your star sign is Aries (that is, if you were born between 21 March and 19 April), you are typically said to be energetic, independent, impulsive and dynamic. Often seen as a risk-taker who is full of new ideas, ambitions and schemes, you're definitely not a wallflower – no sitting alone in the corner of the local dance hall for you. On the down side you might also be intolerant, hasty, arrogant or even domineering. Famous Arians have included Vincent van Gogh, Adolf Hitler ('Who, me, domineering?'), Elton John, Russell Crowe, Sarah Jessica Parker, Samuel Beckett, Billie Holiday and Marcel Marceau.

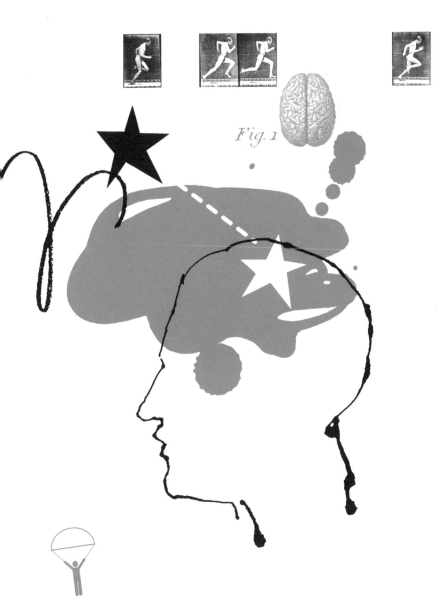

Fig. 1

. . . .

Did you know?

During his time as US president (1913–21),
Woodrow Wilson grazed a flock of sheep on
the lawn of the White House. He sold their wool
to help raise funds for the Red Cross during
World War I.

. . . .

The battle for sheep

Sheep figured in the US War of independence. In 1699, the English government passed the *Wool Act* in an attempt to protect the local wool industry. The *Act* limited production of wool in Ireland and prohibited the importation of wool from the thriving mills in the American colonies. The colonies' wool industry was also seen as a threat to England's dominance in the supply of wool back to those same colonies, and it was eventually banned altogether. Such restrictions, which were also systematically imposed on other industries in the would-be United States, contributed in part to the colonies fighting for independence from the mother country.

Sheep also played a major part in Scotland's notorious Highland Clearances. Beginning in the 1750s, and spurred on in large part by England's expanding agricultural industry, small tenants were evicted by landowners from their Highland homes to make way for more profitable sheep farming. Soaring wool prices saw greed increase, which in turn led to the often brutal removal of people from their fertile interior lands to poor quality coastal areas, a shift that mostly changed their way of life forever as they struggled to adjust to new and unfamiliar ways of earning a living. The often-absent landlords then brought in southern sheep farmers with both capital and experience.

The heavy wool-yielding Cheviot and Blackface sheep were introduced and grazed on land which had previously supported many families, and large sheep runs were eventually established. In 1800 the sheep population in the Highlands was an estimated 377,00; within just 80 years this would rise to around 2 million.

In many areas of the Highlands today, sheep are still farmed where people once lived.

On the sheep's back

To create man was a quaint and original idea, but to add the sheep was tautology.

– Mark Twain

A spin through time

The most prolific of the products we gain from sheep, wool is one of the world's most enduring fibres. It is differentiated from hair or fur in that it has overlapping scales, a little like roof tiles, and it is crimped. Both these qualities make it an excellent spinning and felting material.

While it's believed wool has been used by humans since sheep were first domesticated, it would initially have been plucked from the sheep, rather than shorn – ouch! The earliest remnants of fine wool come from the Greek colony of Nymphaeum in the Crimea, and date from the fifth century BC, while the first known woollen cloth dates from 1500 BC and was found preserved in a bog in Denmark.

It is believed the art of spinning wool is as old as the domestication of sheep, and that the fibres were probably first simply twisted together by hand. Later, a spindle was utilized, and then along came the spinning wheel, said to have been invented in India between 500 and 1000 AD. Spinning wheels did not reach Europe, however, until sometime during the medieval era.

. . . .

Did you know?

The Bayeux Tapestry, detailing the story of William the Conqueror's invasion of England in the 11th century, is in fact not a tapestry at all. One of the world's most famous historical records, it is an embroidery made of wool on linen.

. . . .

A fast track to modern times:

- Although the locals had already established their own wool industry, the invading Romans, well acquainted with the ways of wool, established Britain's first 'wool factory' around 50 AD.
- In the 11th century, William the Conqueror brought skilled Flemish weavers across the waters to mainland Britain. And by the 12th century, wool was well on its way to becoming one of Britain's most valuable assets. Weaving became a widespread industry.
- It was soon realized that maximum revenue could be gained through the export of raw wool rather than already woven cloth. Export taxes on wool largely fuelled England's so-called Hundred Years War with France (1338–1453).
- Wool continued to be used in other corners of the globe as it had been for centuries. In the Middle East and the Mediterranean it was used, as it still is, to make cool clothes to combat the region's heat. And kilims, the simple, flat woven carpets said to have been made for over 4000 years, continued to be produced and utilized as both rugs and portable floors.
- By the end of the 18th century, the Industrial Revolution changed the nature of the English wool industry, seeing textile industries relocated from the home hearth to the factory floor. By that time, too, Britain had over 300 laws pertaining to the wool and sheep industries.

Wool de moderne

Today, wool is commercially produced in many countries around the world. But the industry leader is without doubt Australia. From an initially modest holding of only 13 Merinos, transported to the fledgling colonial outpost in 1797, the country now shears around 107 million sheep per year – of which over 84 per cent are Merinos, with the Peppin Merino being the most prominent and highly regarded breed – and provides the rest of world with around 24 per cent of its greasy wool needs. (Oh, the greasy part? That's how raw, unwashed wool, particularly oily when straight from the sheep's back, is described.) In 2004–05, a total of 475 million kilograms (over 1 billion lb) of greasy wool was shorn from Aussie sheep.

Australia exports wool to 52 countries, with China being the biggest customer, annually purchasing around 170 million kilograms (374 million lb) of the fluffy stuff. And of all the wool used to make clothing around the world in 2004, a huge 51 per cent of it came from Down Under. Though no match for Australia, the other countries rounding out the top five world wool producers are, in order: New Zealand, China, the former Soviet Union and Argentina.

The ultimate all-rounder

Remember that one student at school who could seemingly do it all? Science, art, debating, swimming, drama – for her, nothing was out of reach and, much to the annoyance of most other kids, she excelled at everything. Wool's a bit like that.

It's no accident that wool has been used for so long by us humans. The fibre's amazing characteristics have ensured its continuing place in wardrobes and on floors across the globe. For starters, wool's insulating properties work both ways; not only will it keep you toasty in cold weather, it can also keep you cool in the heat. Both the nomadic Bedouins and the Tuaregs of western and northern Africa use finely woven woollen garments as a guard against the regions fiercely high temperatures. Wool is
also water repellent and fire resistant, is hard wearing and elastic, and takes well to dyeing (particularly before it is woven – the phrase 'dyed in the wool', which describes a person who has deeply held, unshakeable ideas or beliefs, relates to wool that is dyed at this stage).

The mighty Merino looms large in the international wool industry. Renowned for its heavy fleece of fine quality wool, the Merino is the most numerous breed of sheep in the world today.

Merino sheep produce mostly white wool, and although there are occasional variants on this, coloured sheep are generally removed from the flock as their wool is less commercially valued – it cannot be dyed as successfully as white wool. Coloured Merino wool is often used by home spinners and those in cottage industries.

So how is the value of today's wool judged?
- **Fibre diameter** The smaller the fibre diameter, or micron measurement, the more valued the wool.
- **Staple length and strength** The longer and stronger, the better.
- **Colour** Here, whiteness and brightness are highly valued.
- **Yield** This indicates the amount of clean wool left after dirt, grease and other material has been cleaned from the fleece.

And how is it measured?
- **GFW or greasy fleece weight,** indicates the weight of the wool while – that's right – still greasy.
- **Micron** is the term used to describe the diameter of a fibre. One micron constitutes one-millionth of one metre (39 ⅓ in), and is the most scientific method of grading wool.
- **BWT** indicates the body weight of an adult sheep.

And categorised?
- **Superfine** The most valued category, with fibres measuring equal to or less than 19.5 microns, with a GFW of 3–4 kilograms (7–9 lb) and a BWT of 35–40 kilograms (77–88 lb).

- **Fine** Wool measuring 18.6–20.5 microns, having a GFW between 3 and 5 kilograms (6 ½–11 lb) and a BWT of 35–40 kilograms (77–88 lb).
- **Medium** Wool measuring 20.6–22.5 microns, with a GFW of 4–6 kilograms (9–13 lb) and a BWT of 40–50 kilograms (88–110 lb).
- **Strong** Wool having the greatest fibre diameter, measuring at least or greater than 22.6 microns, with a GFW of 5–7 kilograms (11–15 ½ lb) and a BWT of 45–55 kilograms (99–121 lb).

. . . .

Did you know?

The most expensive bale of wool ever sold was produced by Australian woolgrower Barry Walker. In 2004 he produced a 91-kilogram (200 lb) bale of wool, the finest in the world at a tiny 11.8 microns, which sold for AU$227,000 (£93,500) or around AU$2500 (£1030) per kilogram.

. . . .

In the outback shearing shed

Although technological developments have brought sweeping changes to many a workplace, a typical shearing shed is still a harsh working environment – shearing is classified as the most dangerous agricultural occupation in Australia today, and is regarded as one of the heaviest kinds of work in the developed world.

On any working day in a shed you'll find a team of shearers who head a production line of some of the hardest working people you'll find anywhere. Once they've shorn off the fleece, it is then handled by the first shed-hand, or rouseabout, who picks it up from the shearing board and spreads it onto the wool classing table. The second shed-hand, also known as the wool roller, then removes all the short, dirty and inferior pieces of wool and rolls the fleece, ready for the wool classer to move in, assess the wool's quality and class it. The fleece is then pressed into bales by the imaginatively named wool presser. To meet industry standards, each bale of wool must weigh between 110 and 204 kilograms (242–450 lb), with most averaging 170–190 kilograms (375–419 lb).

Before the introduction of shearing machines in Australia in 1887, sheep were shorn using manual blade shears that looked a little like a giant pair of scissors. By the turn of the century most sheds were utilizing the shearing machine that had been invented by Frederick York Wolseley. Today, the majority of sheep are shorn annually, using a mechanically driven handpiece still based on Wolselely's original design. While the average number of sheep an Australian shearer will shear in one day is around 120 to 140, a very good shearer, known as a gun shearer, might process 200 sheep in a day. And that means hauling about 10 tonnes of squirming animal every day, while the shed-hands would lug almost 1 tonne of wool from this one shearer alone.

Did you know?

- The record for the fastest shearing of one sheep is held by Australian Dwayne Black, who in April 2005 shore his sheep in just 45.41 seconds.
- The record for the most sheep manually shorn in 8 hours is held by Janos Marton of Hungary, who denuded a total of 50 sheep in 2003.
- In 1988, in what they termed a 'shearathon', New Zealanders Keith Wilson and Alan MacDonald claimed their place in the record books by machine-shearing the most sheep in 24 hours – 2200 sheep in all.

Warp and weft

It is difficult to pinpoint the moment in history when the craft of weaving cloth began – few remnants of cloth have survived. But it is understood to have been with us for thousands of years, with basket weaving being the first method of taking a material, placing it at right angles and interleaving it to form a woven structure.

Weaving looms have been utilized since primitive times, the basic principle behind the loom remains the same today. The first looms are said to have been hung from a tree branch. The vertical, or warp, fibres would be tied over the branch, and the weft fibres would be interwoven horizontally to the warp. Later, a stick was introduced which served as the shuttle, carrying the weft fibres through the warp.

Various kinds of loom developed through time, one of these being the backstrap loom. This kind of loom ran parallel to the ground; one end was perhaps anchored to a tree trunk, with the other tied around the weaver's back via a strap. The weaver could then simply lean against the pull of the warp to create tension. Backstrap looms are still used today by indigenous weavers in Central America, the USA and Mexico.

Many a visitor to Turkey has come to the end of their holiday wondering just how they are going to transport their newly acquired carpet home. Long-lasting and beautifully intricate in design, for many the Turkish carpet represents the pinnacle of the weaving craft. The first Turkish carpets date from the 11th and 12th centuries, and the first known carpet workshop, created to service the court, was established during the Ottoman Empire. The carpets began to find their way into European castles and homes as early as the 14th century, via merchants in Florence and Genoa.

Two kinds of knotting are mostly used – the Turkish or double knot, and the Persian or single knot, each with its own particular characteristics. Carpets made using the Turkish knot will have a firmer weave and be more durable while those made with the Persian knot are often more intricate in design. And although to many a tourist a Turkish carpet may simply look beautiful, the motifs used in the various designs have a language of their own. A small prayer rug, for example, might feature a motif that conveys the grief of losing a child; it is seen that working on such a carpet may be therapeutic for the weaver. Other motifs depict joy and happiness, some communicate the marital status of the weaver, while others still are a way for a woman to communicate with her descendents, who will inherit the carpet after her death.

Harris Tweed

The world-famous coarse woven cloth known as tweed, traditionally produced on the islands of the Outer Hebrides off Scotland's northwestern coast – Harris, Lewis, Uist and Barra – has today become inextricably associated with English country style. For generations the cloth has been handmade, and when the Industrial Revolution swept through Scotland the weavers of the Outer Hebrides resisted change and kept up their traditional methods of creating their product entirely by hand.

Until the mid-19th century Harris Tweed was known only to those Hebrideans who made it for themselves, their families and the local market. Then in 1846 one Lady Dunmore had a tartan made by the Harris weavers; she loved the resulting product so much she decided to investigate a potential broader market for the tweed, and it is due to her efforts that a wider trade in the cloth around the United Kingdom was initially established.

At the turn of the century the Harris Tweed industry began taking off in earnest, with the introduction of spinning machinery and the more efficient fly-shuttle looms, to replace the older, smaller looms. Importantly at this time, too, those in the industry moved to trademark Harris Tweed, so as to assure customers they were purchasing the genuine cloth, not the imitation mill-spun tweed that was emerging as a competitor.

If you are purchasing a garment made of certified Harris Tweed today, what you get is guaranteed to be made from wool both spun and dyed in the Outer Hebrides. It will have been hand woven by the islanders within their homes and then finished within the islands.

. . . .

Did you know?

The production of wool was so central to the fledgling US economy, that in the mid-1600s the General Court of Massachusetts passed a law requiring young men to learn to spin and weave.

. . . .

An ugg-ly rise to fame

Who would have thought? What was once the humble footwear of Aussie surfers has, in recent years, made its way onto the manicured feet of Hollywood's rich and famous, staking its claim as one of the most surprising fashion trends in history.

The ugg boot is said to have begun life in the early 20th century, when World War I pilots wore fleece-lined boots they called 'fug boots'. In the 1960s the style of boot was adopted by surfers as a way to keep their toes toasty between waves. And although since that time many non-surfing Australians have taken up the ugg boot for comfort and warmth, the phenomenon was largely confined within the nation's borders. Until the 1990s, that is, when Pamela Anderson began sporting a pair on the set of *Baywatch*.

Since that time, the woolly footwear has turned up on the hoofs of such folk as Kate Hudson, Paris Hilton, Gwyneth Paltrow and Russell Crowe, and the craze for uggs spread rapidly across North America thanks both to celebrity endorsers and to the US-owned company Ugg Australia. Uggs were also featured on an episode of the U.S. television hit show *Oprah*, titled 'Oprah's Favorite Things', in 2005. That same year, Ugg Australia's Art & Sole held a charity auction

to raise funds for a children's hospital. And the items to be auctioned off? Ugg boots designed by Reese Witherspoon, Teri Hatcher, Pierce Brosnan, Sarah McLachlan, Brooke Shields, Donald Trump and more. It seemed nothing could stop the rise of the ugg.

But as quickly as fashion trends take off they fade away. Or are badgered into an underground existence. As reported in the *Wall Street Journal* in December 2005, the same magazines that had only recently hailed the arrival of the ugg have now moved on to other trends, and it seems the popularity of ugg boots, in the United States at least, may well be in decline.

Careful, they might hear you ...

Originally begun by an enterprising Australian, Brian Smith, the very successful American company Ugg Australia was acquired by the much larger Deckers Outdoor Corporation in 1995. And, having claimed the ugg-boot company for itself, Deckers then claimed the term 'ugg', registering the once very Aussie word as a trademark in 25 countries, including Australia.

So just as the ugg's appeal was growing, small-time traders in the boot's country of origin were unable to cash in on the success coming the ugg's way. In fact, producers like the Mortel family of the Hunter Valley, on Australia's east coast, who have been making uggs and calling them just that since the 1950s, couldn't even use the term that they claim to have come up with in the first place, lest they face litigation. And hell hath no fury like a giant corporation scorned: Deckers also threatened to sue Australia's *Macquarie Dictionary*, for including in their tome a definition of 'ugg' but failing to acknowledge that the word was now an American trademark.

But just as the fortunes of the ugg boot in the United States appear to have turned, so have those of Deckers in what many see as a petty war. A group of small Australian 'sheepskin boot' producers banded together to take legal action against the American company's trademarking of the word 'ugg'. In early 2006 the Australian national trademark regulator ruled that 'ugh' boot 'ugg' or 'ug' could indeed be seen as generic Australian terms, and that these terms should be removed from the Australian register of trademarks. So once again Australian companies big and small can rightly call their products what they always have, without fear of being taken to court.

Did you know?
In the United Kingdom's House of Lords, the Lord Chancellor sits on a bale of wool, as a display of the historical importance of the fibre to the country's economy.

On your plate

Mutton has formed part of our diet since sheep were first domesticated. And it's been a staple in countries the world over for centuries: in much of northern Africa, where the majority of the population are Muslims and therefore prohibited from eating pork, mutton and lamb continue to be the most widely consumed meats; in China it was particularly popular in meals served at the Imperial Palace during the Ming Dynasty, and dishes included: stir-fried tripe; roast mutton with soy sauce, sesame oil and shallots; and steamed mutton and walnuts wrapped in mulberry leaves. Mongolian hotpot, featuring mutton as the principal ingredient, dates from the rule of Genghis Khan and continues to be a popular dish on Chinese restaurant menus around the globe.

Lamb is meat that comes from an animal up to one year old, with prime lamb – said to be the tastiest and most tender meat – coming from an animal less than eight months old. As with most meats, the best quality will likely come from your local butcher rather than the supermarket. It's here that you can also ask for advice on which cuts are best suited to your cooking needs and for the butcher to cut the meat to your requirements. The leg of the lamb offers the best eating, and can be cut into steaks or a boned roast, or diced for kebabs. The eye of loin offers butterfly steaks and loin chops, while the loin will give you a rack of lamb or cutlets. If you're looking to reduce the amount of fat intake in your diet, looked for Frenched cutlets, which have already had the fat cut off them. As well as these more popular cuts there are many others, including:

- **shoulder, oyster cut**
- **square cut shoulder**
- **leg, chump on, aitch bone removed**
- **leg, chump off, shank off**

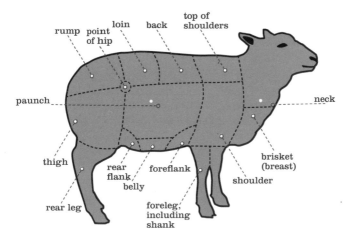

rump point loin back top of shoulders
of hip

paunch

neck

thigh

rear flank

foreflank

brisket (breast)

belly

shoulder

rear leg

foreleg, including shank

- lamb telescoped carcass
- neck
- breast and flap
- hindshank
- trunk meat cuts.

In recent decades in the West, mutton has been nudged from the spotlight by the young upstart, lamb, the production of which now far outstrips that of mutton in countries such as Australia and the United Kingdom. Yet in the United Kingdom there is a push to reassert mutton's place at the dinner table. In 2004 the National Sheep Association launched its first Festival of Mutton, championed by HRH the Prince of Wales, as a way of promoting mutton once again as a tasty British meat and to help find new markets for sheep farmers and their produce. The festival aims to bring to public attention restaurants that are putting mutton back on their menus, along with farms, producers, suppliers and butchers who are once more raising and stocking the meat.

World's best ever roast leg of lamb

Preparation: 20 minutes
Cooking: 1 ¾ hours, serves 6

2 kg (4 lb) leg of lamb
2 cloves garlic, cut into thin slivers
2 tablespoons fresh rosemary sprigs
2 teaspoons oil

Mint sauce
½ cup (10 g/¼ oz) fresh mint leaves
3 tablespoons caster sugar
¾ cup (185 ml/6 fl oz) malt vinegar

Preheat the oven to 180°C (350°F/ Gas 4). Using a small,
sharp knife, cut small slits all over the lamb. Push both the
garlic and rosemary into the slits. Brush the lamb with the
oil and sprinkle with salt and black pepper. Place on a rack in
a baking dish. Add ½ cup (125 ml/4 fl oz) water to the dish.
Roast the lamb for about 1 ½ hours for medium, or 1 ¾ hours
for well done, basting often with the pan juices.

Keep the lamb warm and allow to rest for 10–15 minutes
before serving.

Mint sauce
To make the mint sauce, place the mint leaves on a chopping
board and sprinkle with 1 tablespoon of the caster sugar,
then finely chop the mint. Place in a bowl and add the
remaining caster sugar. Cover with 3 tablespoons of boiling
water and stir until the sugar has dissolved. Stir in the
vinegar. Serve with the roast lamb.

World's best ever Lancashire hotpot

Preparation: 20 minutes
Cooking: 2 ¼ hours, serves 8

8 forequarter lamb chops, cut 2.5 cm (1 in) thick
4 lamb kidneys, cut in quarters, cores removed
¼ cup (30 g/1 oz) plain flour
50 g (1¾ oz) butter
2 large onions, sliced
2 celery sticks, chopped
1 large carrot, chopped
4 potatoes, peeled and thinly sliced
1 ¾ cups (440 ml/14 fl oz) chicken or beef stock
200 g (6 ½ oz) button mushrooms, sliced
2 teaspoons chopped fresh thyme
1 tablespoon Worcestershire sauce

Preheat the oven to 160°C (315°F/ Gas 2–3). Trim the meat
of excess fat and sinew, and toss the chops and kidneys in
the flour, shaking off any excess. Heat the butter in a large
frying pan and quickly brown the chops, in batches, on
both sides. Remove the chops from the pan and brown the
kidneys. Remove the kidneys and add the onion, celery and
carrot to the pan and cook until the carrot begins to brown.
Brush a large casserole dish with a little melted butter or
oil. Layer half the potato slices in the base of the casserole
dish and place the chops and kidneys on top. Then layer the
onion, celery and carrot on top.

Sprinkle the remaining flour over the base of the pan
and cook, stirring, until brown. Remove from the heat,
pour in a little stock and stir until smooth. Add the rest of
the stock gradually, return to the heat and stir until the
mixture boils and thickens. Add the mushrooms, thyme and

Worcestershire sauce and season to taste with salt and black pepper. Reduce the heat and leave to simmer for 10 minutes. Pour into the casserole dish.

Layer the remaining potato over the top of the casserole, to cover the meat and vegetables. Bake, covered, for 1 ¼ hours. Uncover and cook for a further 30 minutes, or until the potatoes are brown.

World's best ever Irish stew

Preparation: 20 minutes
Cooking: 2 ½ hours, serves 4

1 kg (2 lb) lamb neck chops
3 tablespoons plain flour
4 onions, sliced
8 potatoes (about 2 kg/4 lb), quartered
¼ cup (15 g/½ oz) chopped fresh parsley

Preheat the oven to 160°C (315°F/ Gas 2–3). Trim the lamb chops of fat and sinew and toss in the flour, shaking off any excess. Place the chops in a 2-litre ovenproof dish with the onion and 2 cups (500 ml/16 fl oz) of water. Bake, covered, for 1 ½ hours.

Add the potato and bake, covered, for a further 1 hour, or until the meat and potatoes are tender. Stir in the parsley just before serving.

Note: Traditionally Irish stew was made using mutton. This is a modernized version using lamb.

World's best ever lamb's fry and bacon

Preparation: 20 minutes
Cooking: 15 minutes, serves 4

500 g (1 lb) lamb's fry (lamb's liver)
plain flour, to dust
30 g (1 oz) butter
1 tablespoon oil
2 onions, sliced
4 rashers bacon, chopped
1 ½ cups (375 ml/12 fl oz) chicken stock
1 teaspoon Worcestershire sauce

Peel off and discard the outer membrane from the lamb's fry, then cut the lamb's fry into 1 cm (½ inch) slices. Put some of the flour in a shallow bowl and season with salt and pepper. Dust the lamb's fry with the seasoned flour and shake off any excess, reserving 2 tablespoons.

Heat the butter and oil in a frying pan and cook the onion over a medium heat for 3 minutes, or until golden. Remove the onion from the pan and set aside. Add the bacon to the pan and cook over a medium heat until brown. Remove from the pan and set aside.

Add the lamb's fry to the pan and cook over a high heat for 1 minute on each side, or until lightly browned. Return the onion to the pan. Blend the chicken stock, Worcestershire sauce and reserved flour in a jug. Pour into the pan and stir until the mixture boils and thickens. Reduce the heat to low and simmer for 3 minutes, or until the lamb's fry is tender. Stir in the cooked bacon. Serve immediately with toast.

Note: Lamb's fry should be pink in the centre when cooked. Do not be tempted to overcook it or it will become tough.

Mutton in beer, medieval style

From the pages of history comes this recipe for mutton, from the 1572 book *A Proper Newe Book of Cokerye*.

1 kg (2 lb) leg of mutton
500 ml (1 pint) dark beer or ale
2 onions, thinly sliced
1 teaspoon salt
pepper to taste
2 tablespoons butter

Bone the mutton, trimming off any skin or excess fat. Cut into thin slices across the grain. Place in a heavy pan with the beer and onions, cover and simmer on the stove for 1 hour. Add the salt, pepper and butter and continue simmering for 30 minutes or until tender. Serve with fingers of fried bread.

What teams wonderfully with lamb?

Rosemary, thyme or garlic; roast potatoes, honey-glazed carrots or eggplant; Moroccan-inspired flavours – cardamom, cumin, cinnamon, turmeric, preserved lemons; a mix of Indian spices – garam masala, curry, ginger, cloves, mustard seed, bay leaves, coriander seed; Greek style – with yoghurt, barbecued on skewers, with a white wine, fetta and tomato sauce.

Don't throw it away

The famed Scottish dish of haggis utilizes parts of a sheep that might otherwise be left for scraps. Though its origins are unclear, haggis has throughout its history been made by the lower classes, who could not afford to waste any part of the animal. The meal, however – often viewed with horror by those not accustomed to eating offal – underwent a transformation at some stage and even became elevated to royal status: Queen Victoria herself, who adored Scotland and everything about it, is said to have enjoyed the occasional meal of haggis. Haggis is now considered the national dish of Scotland and is readily available in UK supermarkets – you can even buy a deep-fried version at the local fish and chip shop.

A typical haggis consists of what is known as sheep's pluck (lungs, heart and liver), minced together with oatmeal for bulk, suet, salt, spices and stock. This is then stuffed into the sheep's stomach, or maw, boiled up and traditionally served with tatties (potatoes) and neeps (turnips).

The haggis has certainly travelled a long way from its humble, 'waste not, want not' origins – and never more so than the distances it is thrown in the internationally popular sport of haggis hurling. The tradition that started the sport is said to have originated during the clan wars, when the women of Auchnaclory would throw a haggis lunch across the Dromach River to their hungry husbands. To revive this long-lost tradition, in 1977 one Robin Dunseath placed an advertisement in a local Edinburgh newspaper, stating that the national sport would be showcased at the upcoming Gathering of the Clans. Apparently hundreds of people turned up to watch and participate, and from there the tradition was exported to countries such as the United States, Canada and Australia. And it took off to such a degree that Dunseath went on to set up the World Haggis Hurling Association. He also wrote the guide *The Complete Haggis Hurler*, which details the rules, regulations and history of the sport.

Sound like a heap of old sheep's guts to you? It is. While it's true the 1977 Gathering of the Clans was the place where haggis hurling was 'reborn', and the World Haggis Hurling Association was indeed set up by Dunseath, the former PR man did so as a joke, fabricating the entire sport, its history and its rules. In fact, the 'sport' took off to such an extent that it would be 20 years before Dunseath 'fessed up to his role in its new-found fame. In his defence, however, it must be noted that all proceeds raised by the World Haggis Hurling Association were, to salve Dunseath's conscience, donated to charity.

Despite the sport's false start, many still fling their haggis with barely concealed enthusiasm in competitions around the globe. In today's contests the gullible haggis hurler usually stands on top of a whiskey half-barrel rather than a fabled river bank. Alan Pettigrew is the current haggis hurling world record holder, and has been for over 21 years. In August 1984 he broke the record by throwing a 680-gram haggis a distance of 55.1 metres (in Dunseath's 'traditional' rules, all measurements must be recorded in the imperial fashion, so that'd be a 1 lb 8 oz haggis thrown 180 ft 10 in). And, for the record, Pettigrew still believes his chosen sport comes from a long line of hurlers, not from the mind of a spin doctor.

'Gie her a haggis!'

Ever year around the world, Burns Suppers pay tribute to one of Scotland's best-loved poets, Robbie Burns. Traditionally held close to the poet's birthday of 25 January, it is believed that the ritual of the Burns Supper was begun a few years after the poet's death in 1796, by close friends wanting to honour and commemorate his life.

There is a traditional format for a Burns Supper, which has remained essentially unchanged since the first celebration over two hundred years ago. The evening opens with an address by the supper's chairperson and the saying of grace (the Selkirk Grace). Next comes an invitation for those gathered to receive the haggis, to which Burns paid tribute in his 'Address to a Haggis' in 1786. The haggis is then carried into the room, heralded by a piper, and the chairperson begins to recite Burns' difficult-to-follow poem, the final verses of which are as follows:

But mark the Rustic, haggis-fed,
The trembling earth resounds his tread.
Clap in his walie nieve a blade,
He'll make it whistle;
An' legs, an' arms, an' heads will sned,
Like taps o' thrissle.

Ye pow'rs wha mak mankind your care,
And dish them out their bill o' fare,
Auld Scotland wants nae skinking ware
That jaups in luggies;
But, if ye wish her gratefu' prayer,
Gie her a haggis!

Blessed are the cheesemakers

Sheep's milk may contribute a mere 1.3 per cent of the total world milk production, compared to 84 per cent for cow's milk, but for many people it is a very important product. Richer than cow's milk in calcium, magnesium and a variety of vitamins, sheep's milk is also more easily digested than cow's milk and is a good dairy alternative for those allergic to the milk of cows, including children.

In many European countries sheep's milk is used to make internationally renowned cheeses, and the work of the cheesemaker has become a fine art, the skills of which are passed down through the generations. In France, roquefort cheese has been made, so the legend goes, since Roman times. In 1411 Charles II bestowed sole roquefort-making rights on the village of Roquefort; today, cheese must still be aged in the caves around Mont Combalou, close to the village, in order to be officially classed as roquefort. The blue veins of the cheese were originally a by-product of its ageing in the limestone caves, but today the mould is injected into the cheeses to produce the same result.

Fetta, that wonderfully salty, crumbly cheese beloved of Greeks, is made with sheep's milk, though lesser versions are sometimes made with cow's milk. Believed by some to be the cheese mentioned in Homer's *Odyssey*, fetta is traditionally made by adding rennet to the milk, to form curd. The curd is then hung in cloth bag to dry for a few hours, after which it is preserved with salt and aged in brine.

Ricotta cheese is one of the most versatile of cheeses made from sheep's milk. Originating in Italy, ricotta cheese is not strictly a cheese, but is made from the skimmed-off leftover whey which results from making other cheeses. Ricotta can be used in both sweet and savoury dishes – it can be used to stuff pasta such as cannelloni or ravioli, made into delicious desserts like cheesecake, and used to make quiches and sauces.

On your face

Waste not, want not. Lanolin, the grease contained in raw wool, is utilized in a number of ways. Extracted during the process of scouring the wool, lanolin is used in products as varied as motor oils, inks, adhesive tape and cosmetics.

Skin care products containing lanolin include lip balm and moisturiser – lanolin's consistency is said to protect our skin in the same way it protects the sheep's wool from the effects of the environment. Lanolin is believed to have been used since the times of the ancient Greeks. It contains cholesterol and natural fats which make it an ideal match for moisturising human skin.

Other sheeply skin care products are made from the less glamorous sounding sheep placenta. Companies selling creams, lotions and potions that include the product sing placenta's praises; their claims range from assuring their products' ability to smooth skin and improve its elasticity to its miraculous ability to not only halt but actually reverse the ageing process. Hmmm.

Did you know?

Sheep by-products are used in the production of, among other things: brake fluid, wallpaper, floor wax, photographic film, toothbrushes, baseballs and tennis balls, crayons, shampoo and conditioner, antifreeze, insulation, piano keys and billiard table covers.

On the nose

One product from sheep you probably won't be seeing, or
smelling, on the catwalks next season is methane. One of the
three most potent gases believed to be causing accelerated
global warming, methane is a by-product of our ovine friends,
amongst others, the result of a diet high in the condensed
tannins found in plants. And one country where the
population can genuinely blame someone else for somewhat
unsocial odours is New Zealand. With such a high population
of livestock, the country is unique, in that up to an estimated
50 per cent of their total greenhouse gas emissions – and a
whopping 90 per cent of methane emissions – are caused by
their livestock.

What part does the humble NZ sheep play in all of this? Take,
for example, a flock of 300 sheep. Every year, the flock will
produce approximately 21,900,000 litres of methane (pew!)
and around 81,125,000 litres of CO_2. With a total population
of around 40 million sheep (and don't forget the 10 million
cows and various other ruminants) that's a national issue
that's definitely on the nose. So great is the problem that
in 2003 the New Zealand federal government proposed
introducing a 'flatulence tax' on farmers. The proposal was
part of the government's attempts to meet their commitments
to the Kyoto Protocol. It didn't pass.

Scientific sheep

Trouble is, sheep are very dim.
– Monty Python, 'Flying sheep' sketch

Are sheep intelligent?

Haven't we met before? While they might not be too good with names, sheep have a good memory for faces. CSIRO researchers in Armidale, New South Wales, recently tested both the intelligence and learning abilities of sheep, and came up with some interesting findings. When tested in a complex maze similar – though no doubt larger – to that used for rats, the sheep were found not only to possess outstanding spatial memories, they also displayed an ability to learn from experience. And they can retain this new-found knowledge for up to six weeks. Not bad for a creature that's usually considered a little on the slow side.

So what does this mean for the future of these fleecy creatures? The CSIRO is hoping its findings will eventually make for happier sheep; they aim to identify smarter sheep who'll be easier to look after and will be more readily adaptable to changing farming systems.

If being able to find your way through a complex maze isn't impressive enough for you, how about a sheep that can recognize its buddies? Research at the Cambridge Babraham Institute in the United Kingdom has uncovered the somewhat surprising fact that sheep can identify individual human faces and remember them for an impressive two years. So elephants may not be the only animals who don't forget. Even more astounding, perhaps, is the discovery that they can recognize their 'friends', sheep they're already acquainted with.

What's more, they may also experience that warm and fuzzy feeling of mateship. When let into a maze containing photos of both familiar and unfamiliar sheep faces, it was observed that 80 per cent of the sheep chose to stick to the route through the maze peopled by their mates. In fact, when shown photos of familiar ovine faces their brains apparently process the visual information in the same way as ours do when looking at happy snaps of our friends from last Saturday's dinner party.

While these findings might lead you to ponder what would happen if sheep put their intelligence and memory to use, maybe to break out of their paddock and seek revenge on their keepers, just remember – they are still, well, a sheepish lot.

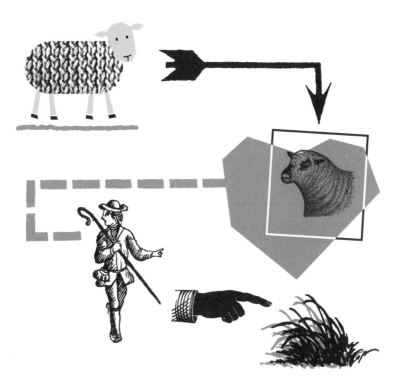

An Australian sheep farmer, a Mr Johnston from Yungaburra in Queensland, claims to have seen intelligence coupled with compassion at work in his flock. While out mustering one day, Mr Johnston came upon a ewe and her lamb. Being totally blind, the ewe relied on its offspring both to direct it and keep it from harm – he even saw the lamb taking the lead and jumping over fallen logs first in order to safely guide its mother.

Sheep have been witnessed displaying a rather nifty way to get themselves over cattle grids. In the Yorkshire village of Marsden, cattle grids were installed in an effort to curb the village's ravenous sheep, who had been spotted munching on local gardens, the village bowling green and even grass in the graveyard. The grids have proved no match for the ingenious sheep, however, who have taught themselves to simply roll across the 3-metre (8 ft) expanse.

. . . .

Did you know?

'The sixth sick sheik's sixth sheep's sick' is thought to be the most difficult tongue twister in the English language.

. . . .

Dolly the sheep
is born

In the lab

Arguably the most famous ruminant around, Dolly the
sheep rose to prominence in early 1997 as the first mammal
successfully cloned from adult cells. Born on 5 July 1996 at
Scotland's Roslin Institute from a six-year-old ewe, Dolly was
a Finn Dorset and went on to have four healthy, naturally
conceived lambs.

Dolly was not the first sheep the Roslin team had cloned;
a year before Dolly's birth, twins Megan and Morag had been
cloned from cells that had been grown in a lab for several
weeks. But unlike Dolly, the cells used were embryonic,
not adult, and so Megan and Morag did not generate much
interest.

What was so special about Dolly's cloning? As opposed to cloning from stem cells, which are essentially 'un-programmed' and can become any number of different cells within the body (hair, skin, muscle, lung etc), a fully developed adult cell, such as that from which Dolly was cloned, is already programmed. The challenge for the Roslin scientists, then, was to de-program the cell in order to produce a successful clone. To do this, they de-nucleated the cell and then inserted a nucleus from a parent cell.

Dolly's cloning brought up a number of ethical issues and prompted much public debate. Is cloning animals for human benefit ethically right? Is it natural to clone an animal? Will this lead to the cloning of humans and what would be the possible implications?

While Dolly died at the relatively early age of six, her post-mortem found no evidence to suggest her death was linked to her status as a clone. During her life, however, she showed signs of early ageing, and it is believed this may have been linked to her genes coming from an already six-year-old ewe, thus raising the possibility that when Dolly was born she was actually, genetically speaking, six years old herself.

Did you know?
Dolly the sheep may be dead but she's not forgotten – and she still hasn't left the limelight. After her death the taxidermists moved in and she is now on display, waiting for you to call upon her, at the Royal Museum in Edinburgh, Scotland.

Perhaps not as famous as Dolly, Tracy led the charge of sheep into the realm of transgenics. In 1990, Scottish-based biotechnology company PPL Therapeutics transferred genetic material from one species, humans, to another, sheep – the result being Tracy. This perhaps sounds more Frankenstein in nature than it really is; the process involves modifying a single cell from a sheep, so that it includes the human gene for the protein alpha-1 antitrypsin (AAT). What they then found was that Tracy's milk contained AAT, a human protein that can be used to help treat cystic fibrosis.

While AAT can be derived from blood plasma, factors such as high cost, possible contamination and inefficiency in extracting AAT make the method developed by PPL a much more enticing prospect. And it's hoped that such research could lead to the discovery of techniques to produce other disease-fighting drugs.

Researchers at New Zealand's AgResearch have developed a breed of sheep that is essentially free of wool on its face, belly, legs and bum. The breed was originally developed to eliminate the need for mulesing, where folds of skin are removed to prevent the maggots of blowflies infesting the skin (known as sheep strike or fly strike). It was found, however, that the breed also offers the possibility of increased productivity in the shearing shed, saving time with both shearing and handling of the fleeces. The scientists at AgResearch are hoping that commercial producers will be able to breed these sheep themselves within the next five years.

While sheep are a little more fussy in their eating habits than their close relatives, goats, they will apparently happily chow down on a wholesome meal of tender, tasty weeds. And in the

United States this preference for poorer pastures is being put to good use.

Researchers at the Sheep Experiment Station in Dubois, Idaho, have discovered that some sheep love the taste of knapweed, an introduced European species that is causing problems for the local sagebrush. By grazing sheep in the right place at the right time – when the weed is growing and the native grasses aren't – the very invasive knapweed can be controlled, giving the more fragile sagebrush, which the knapweed outperforms, a chance at survival.

Environmental groups are proposing that the government pay farmers to graze their flocks in targeted areas to help reduce the spread and effects of knapweed. The task isn't entirely straightforward, however, as grazing cattle, and sheep, still eat native grasses in the many areas in which they feed, creating perfect conditions for the knapweed to take hold. It is hoped, though, that selective targeted grazing may be the start of a control program.

Sheep urine is known to reduce pollution emissions from vehicle engines. In the United Kingdom, the bus company Stagecoach South are putting this to the test, having last year installed in one of their vehicles a tank containing the very liquid. The waste is sprayed into the exhaust fumes, thus lowering the emission of nitrous oxides. Ammonia in the urea – the refined form of urine which is used in the bus – interacts with the nitrous oxides, converting them into less harmful nitrogen gas and water. This is then expelled as steam.

In an interview with the *Guardian*, Stagecoach South's managing director Andrew Dyer assured the public they would not be required to give samples, nor would there be a sheep riding at the back of the bus with them.

No shrinking violet

Australia's CSIRO and the Western Australian Department of Agriculture have collaborated on a project where they have might just have found the way to maintain the perfect fit of your woollens. Their studies have uncovered the fact that wool shrinkage, also known as felting, is a heritable trait. So, the wool from some sheep is more prone to shrinkage than the wool from others.

The findings of this research will enable farmers to identify those sheep within their flocks that have a low-shrinkage fleece, and then to utilize these particular sheep for further breeding. Not only would increased production of low-felting wool be beneficial to the garment industry – where felting causes problems with the manufacture of both woven and knitted products – it also appears that such wool is less likely to break during spinning, a result of the longer length of its fibres. And it pills less, too.

Corny, tacky, often crude – call them what you will, but there is plethora of ovine jokes out there. Here's a quick grab-bag of fifteen.

1 Why did the lamb call the police? She'd been fleeced.

2 If a philosophical sheep is cloned, what will it search for? The meaning of the eweniverse.

3 What do you get if you cross a sheep with a kangaroo? A woolly jumper.

4 What do you call a sheep without legs? A cloud.

5
'Sheep May Safely Graze' by J.S. Bach has eight bleats to the baaaaa

6 Little Bo Peep slept with her sheep, The sheep was a ram, And now Little Bo Peep is going to have a lamb!

7 A young man grew fed up with modern life and decided to leave the big city to become a shepherd. At first he spent months in the seclusion of the distant mountains alone with his thoughts and sheep. So he moved futher up into the mountains where he found three older shepherds with a big flock of sheep, and asked them to show him the ropes. The shepherds agreed.

The young man spent a week with them. One evening by the fire he asked casually, 'So how do you guys get by with no women around here?' One of the men replied 'Why, with so many sheep around, who needs women?' The youngster shuddered. 'Yuk! How horrible! How can you...?' The three men only smiled and said nothing.

Another week passed and one morning the young man realized that the tension in his groin had grown unbearable. He remembered what the men had said, and looking at the sheep, thought, 'Hmm, why not, after all...? He chose a moment when none of the older shepherds were around, and grabbed one of the nearest sheep. However, the others showed up in a minute, and seeing him with the sheep burst out laughing.

'What? What?' shouted the young man, blushing. 'You told me that's what you did yourselves, didn't you?'

'Yeah, sure. But to choose the ugliest one??!'

8 Why did the sheep get arrested?
She made a ewe turn!

9 What did one cloned sheep say to another?
I am ewe.

10 There were two sheep in a pub, and one of them walked into the baa.

11 Australia: where men are men, and sheep are nervous.

12 A farmer and his wife were lying in bed one evening; she was knitting, he was reading the latest issue of *Farmer's Weekly*. He looked up from the page and said to her, 'Did you know that humans are the only species in which the female achieves orgasm?'

She looked at him wistfully, smiled and replied, 'Oh yeah? Prove it.'

He frowned for a moment, then said, 'Okay.' He then gots up and walked out, leaving his wife with a confused look on her face.

About a half an hour later, he returned all tired and sweaty, proclaiming 'Well I'm sure the cow and sheep didn't, but the way that pig's always squealing, how can I tell?'

13 How do sheep carry their wool? In a baaaaaaaaaaaag.

14 How did the New Zealand navy get lost at sea?
They radioed a ewe-boat.
What happened to the ewe-boat?
Ask the Australians, they rammed it ...

15

A ventriloquist was returning home from a show one evening when his car broke down on a lonely country road. He left his car on foot, in search of help, and eventually came across a farmhouse. He knocked on the front door, explained his plight to the farmer who answered, and the farmer invited the ventriloquist to stay the night at his farm.

Once inside the farmhouse, the ventriloquist was surprised to see that the animals lived side by side with the farmer—there was nothing dividing the farmer's living space from that of his animals. Being a ventriloquist, he decided he'd have a little fun with the old farmer.

'Do you mind if I have a chat with your horse?' he asked.

'The horse don't speak,' the farmer replied.

Ignoring this, the ventriloquist asked the horse, 'So, how's the old guy treating you, then?'. Throwing his voice and pretending to be the horse, the ventriloquist answered, 'Not too bad. He works me very hard but it's nothing a little more hay wouldn't make up for.'

The farmer looked stunned. The ventriloquist then turned his attentions to the farmer's dog, and asked, 'How has your day been?', to which the dog seemingly replied, 'I can't complain. Though I'd appreciate some more meat now and then, especially when he makes me herd sheep all day.'

Again, the farmer took a few steps back in amazement.

Suddenly, a sheep's bleating was heard. 'The other animals have been so talkative,' the ventriloquist mischievously said, 'would you mind if I had a word with your sheep?'

As quick as a shot, the farmer replied,' No bloody way! That ewe tells lies!'

Celebrity sheep

Let me be weak, let me sleep,
and dream of sheep.
– 'And Dream of Sheep', Kate Bush

Art through the ages

The earliest recorded sighting of cave paintings in the Yinshan Mountains, in Inner Mongolia, dates to the fifth century AD, when geologist Li Daoyuan wrote of seeing the primitive art. The paintings are said to date back to four ages: the first, from the late Paleolithic through to the Bronze Age; the next date from the Western and Eastern Han Dynasties (206 BC–220 AD); the third are from the Middle Ages; and fourth are works of the Mongolians, after the Yuan Dynasty (1271–1368). While the paintings number in the thousands, the most numerous are those that depict animals, including sheep and blue sheep.

Other rock art where sheep have cropped up occur in Baja, California, in a cave known as Cueva de Pintada. Here, huge mountain sheep jostle for space along 152 metres (500 ft) of painted walls with human figures, red and black deer and other animals. Though the date of the paintings is uncertain – they are believed to have been painted somewhere between 100 BC and 1300 AD – what is known is that the artists were a local indigenous population who have now disappeared.

As well as cave paintings, there are also numerous petroglyph, or rock carving, sites depicting sheep on the North American continent, including at Moab, Utah; at the Coso Range in the northern Mojave Desert; at Writing-on-Stone Provincial Park, Alberta; and at the Three Rivers Petroglyph site in New Mexico.

It seems that artists in 19th century Britain weren't concerned about the fat content of their lamb. In contrast to today's concerns with reduced fat and lean meat, breeders of times past wanted to produce animals that were bulked up with fatty, marbled meat, much prized at the time. And what better way to put your fat flock on display than to commission a painting of their rotund rumps.

Paintings in the naïve or primitive style were often commissioned by farmers during the 19th century to show off their animals' best assets. One such painting, which depicts five Leicester rams, was executed by J.B. Wood and in 2006 was placed for auction with esteemed UK auction house Bonhams, with an estimated value between AU\$24,680 and \$37,000 (£10,000–15,000).

Painted between 1880 and 1890 at the Brockelsby Station in New South Wales, Tom Roberts' 'Shearing the Rams' depicts a scene of hard work and a tradition that has changed little in Australia in over 100 years. In the scene, shearers and shed-hands are seen hard at work in what must have been a hot, fly-infested and unforgiving environment. While today some criticize such paintings as evoking an overly sentimental picture of the past, 'Shearing the Rams' is considered by many to show a quintessential Aussie country scene. (For more on the work of the Australian shearer, see 'In the outback shearing shed' in the third chapter).

Notorious for the sensational nature of his work, contemporary UK artist Damien Hirst created his piece 'Away from the flock' in 1994. That same year, the work – a dead sheep suspended in a formaldehyde-filled tank – sold for AU\$606,500 (£250,000). While Hirst has many detractors, he has had support from unlikely quarters; sheep farmers have thanked him for raising the profile of the humble lamb, and are quoted as calling him 'a good judge of sheep'. Other Hirst works include 'The physical impossibility of death in the mind of someone living', featuring a preserved shark, and 'This little piggy went to market, this little piggy stayed home', depicting a preserved pig.

Sheep in the nursery

Arguably the most famous sheep in the nursery belonged to Mary. 'Mary had a little lamb' was first published by Bostonian Sarah Hale in 1830 in her collection *Poems for Our Children*. And for such a seemingly innocent nursery rhyme it has stirred some controversy over the years.

The poem was based on the story of Mary Sawyer, a schoolgirl in Sterling, Massachusetts, who used to keep a pet lamb. Mary and her lamb caught the eye of John Roulstone, a college student who visited Mary's school one day with his uncle, the local minister. The day after the visit, Mary herself recalls, Roulstone 'rode across the fields on horseback to the little old schoolhouse and handed me a slip of paper which had written upon it the three original stanzas of the poem ...'

There are two theories on the poem's origin. One upholds Sarah Hale as the author of the entire poem, while the other states that Roulstone, as Mary apparently recalls, wrote the first 12 lines and that Hale added the last 12.

'Mary had a little lamb' was recorded by Thomas Edison in 1877 onto his new invention, the phonograph, making it the first ever audio recording. Edison's close friend Henry Ford, of automobile fame, was apparently so taken with the poem that in 1923 he had what he believed to be the original schoolhouse relocated from Sterling, Massachusetts, to a property he owned in Sudbury, in the same state.

Paul McCartney's post-Beatles band Wings released a version of the nursery rhyme with a new melody in 1972. Such a move begs the question, 'Why?'. Apparently Mary McCartney, the daughter of Paul and wife Linda, liked to hear her name sung when she was a small child, so the recording was made for her. The new release stayed in the charts for 11 weeks and made it, unbelievably, to number 9.

Mary had a little lamb

Mary had a little lamb,
little lamb, little lamb,
Mary had a little lamb,
its fleece was white as snow.
And everywhere that Mary went,
Mary went, Mary went,
and everywhere that Mary went,
the lamb was sure to go.

It followed her to school one day
school one day, school one day,
It followed her to school one day,
which was against the rules.
It made the children laugh and play,
laugh and play, laugh and play,
it made the children laugh and play
to see a lamb at school.

And so the teacher turned it out,
turned it out, turned it out,
and so the teacher turned it out,
but still it lingered near,
and waited patiently about,
patiently about, patiently about,
and waited patiently about
til Mary did appear.

'Why does the lamb love Mary so?'
love Mary so? love Mary so?

'Why does the lamb love Mary so,'
the eager children cry.
'Why, Mary loves the lamb, you know,'
the lamb, you know, the lamb, you know,
'Why, Mary loves the lamb, you know,'
the teacher did reply.

The well-known nursery rhyme 'Baa Baa Black Sheep' was first published in 1744, and is attributed to the elusive Mother Goose (a nom de plume that first appeared in France in the late 17th century and has never been successfully traced to any particular author), while the accompanying tune first appeared in the early 19th century. It is believed that the poem has its origins as a literary backlash against the English Customs Statute of 1285, which authorized the king of the day, Edward I, to collect a tax on all wool imports through every port in the country.

A warning issued by the Birmingham City Council in 1999 recommended the rhyme not be taught at schools for fear it may be seen as racist. The warning was scrapped a year later, however, after black parents labelled it 'ridiculous'. But although Birmingham's parents have seen sense, others apparently haven't. In another part of the United Kingdom, two Oxfordshire nurseries, run by a charity group known as Parents and Children Together (PACT), are teaching children to sing 'Baa Baa Rainbow Sheep'. The group, however, asserts that the move has not been driven by concerns of racism, but rather by a desire to encourage children to use a wider vocabulary.

Baa baa black sheep

Baa baa black sheep,
have you any wool?
Yes sir, yes sir,
three bags full.
One for the master,
one for the dame,
and one for the little boy
who lives down the lane.

For many people the first verse of Little Bo Peep is the only
one with which they are familiar. But it's in the following
verses that the moral to the story uncovers itself. With an
uncertain history – some saying the first verse predates
Victorian times, with the rest being added at a later date –
the rhyme tells of taking responsibility for one's actions, and
facing up to the consequences when you don't do as
you should.

Little Bo Peep

Little Bo Peep has lost her sheep
and doesn't know where to find them.
Leave them alone and they'll come home,
wagging their tails behind them.

Little Bo Peep fell fast asleep
and dreamt she heard them bleating;
But when she awoke, she found it a joke,
for they were all still a-fleeting.

Then up she took her little crook
determined for to find them.
She found them indeed,
but it made her heart bleed,
for they had left their tails behind them.

It happened one day as Bo Peep did stray
into a meadow hard by;
there she espied their tails side by side
all hung on a tree to dry.

She heaved a sigh and wiped her eye,
and over the hillocks went rambling,
and tried what she could,
as a shepherdess should,
to tack each tail to its lambkin.

While shepherds watched their flocks by night ...

The image of the religious faithful as a flock of sheep and Jesus as their shepherd is well known in the Christian tradition. And it was a group of shepherds who famously first sighted the Star of Bethlehem that would lead the Three Wise Men to the site of Jesus' birth. So it's not surprising that sheep are mentioned many times in the Bible – a total of 247 times.

Probably the most famous reference to sheep in the Bible is that in Psalm 23:2, where 'The Lord is thy shepherd' – 'He maketh me to lie down in green pastures; he leadeth me beside still waters'. Other biblical nods to sheep describe them as agile (Psalms 114:4 and 114:6), as being prolific (Ezekiel 36:37, Song of Solomon 4:2), and, as followers of Christ are often cast, as the classically innocent creature needing protection (2 Samuel 24:17). Reflecting the economic importance of the animal at the time the Bible was written, sheep are also mentioned as an integral part of one's personal wealth (Genesis 13:5, Genesis 24:35), and as a valuable source of both meat and milk (1 Kings 1:19, Isaiah 22:13, Isaiah 7: 21–2). One of the more intriguing references to sheep in the Bible occurs in the Old Testament, where Jacob uses what can be viewed as genetic selection to increase in his flock those sheep with particularly desirable physical characteristics.

On film

'There are 40 million sheep in New Zealand ... and they're pissed off.' So begins a 2006 press release by the New Zealand Film Commission, heralding the start of filming on a new NZ film, *Black Sheep*. Billed as a horror comedy, the film takes place on a sheep station, where a genetic engineering experiment goes horribly wrong, turning innocent sheep, as the press release puts it, 'into blood-thirsty killers'.

And what does the film's writer and director, Jonathon King, think of working with sheep? They are apparently easy to work with, though perhaps not quite willing to offer up their most menacing looks for the camera. That's where the animatronics of Weta Workshop, the team behind such films as the *Lord of the Rings* trilogy and *King Kong*, will come in, turning these sheepish innocents into the crazed, out-of-control beasts of King's imagination. Readers eager to see the ruminant rampage on the big screen needn't worry that the film won't make it beyond New Zealand's shores: distribution for the film has already been secured in Britain, Australia, Thailand, Malaysia and Singapore.

It wasn't a sheep but a pig who plaintively but memorably sung his way into celluloid history with 'La, la, la ... la, la, la ...'. But as many a fan of *Babe* (1995) will tell you, the sheep share the limelight in this sweet farmyard fable. Probably the world's only known pig-turned-sheep dog, the orphaned pig won the hearts of sheep and audiences alike with his award-winning turn in the sheep dog trial arena.

'More cheese, Gromit!' From the imaginative and, it would seem, infinitely patient, team at Aardman Animation came 1995's *A Close Shave*, starring the unlikely yet endearing heroes Wallace and Gromit. Although Wallace's new invention, the Knit-O-Matic, can shear a whole flock of sheep in record time, there's a sheep shortage on and the pair find they need to supplement their income, and so take to washing windows. Wallace falls for one of his customers, Wendolene Ramsbottom, the local wool shop owner, whose dog is running a sheep-rustling outfit. Wallace's canine companion, Gromit, is framed for sheep rustling and thrown behind bars. Enter Sean the Sheep, a stray who has come to live with Wallace and Gromit. Convinced of Gromit's innocence, Sean spearheads a plan to bring Gromit home, and with the help of Wallace and some of his fleecy friends frees Gromit from prison.

In Australian verse

Written by revered 'bush' poet A.B. Banjo Paterson in 1895, 'Waltzing Matilda' has become something of an alternative national anthem in Australia. The original poem was written by Paterson, whose day job was as a Sydney solicitor, while visiting Dagworth Station in Queensland. The tune was written by Christina Macpherson, a friend of Paterson's fiancé Sara whom he met at the station, and was adapted from an already existing folk song.

The meaning behind the lyrics has been much debated – some scholars see it as a song of rebellion, acknowledging the conflict at the time between squatters, or station owners, and their shearers, to whom the squatters refused to pay higher wages. What is known for certain, however, is the definitions of some of the now-outdated terms used in the song. A swag, or bed roll, was often regarded by a swagman as his de facto wife, and thus called his 'Matilda'. And jumbuck, or sheep, is thought to be a corruption of 'jump up'.

The lyrics themselves have undergone a few minor changes since Banjo penned the original version – here are the words as they are popularly sung today.

Waltzing Matilda

Once a jolly swagman camped by a billabong
under the shade of a coolibah tree,
and he sang as he watched and
waited til his billy boiled,
'You'll come a-waltzing Matilda with me.'

'Waltzing Matilda, waltzing Matilda,
you'll come a-waltzing Matilda with me.'
And he sang as he watched and
waited til his billy boiled,
'You'll come a-waltzing Matilda with me.'

Down came a jumbuck to drink at that billabong,
up jumped the swagman and
grabbed him with glee.
And he sang as he shoved that
jumbuck in his tuckerbag,
'You'll come a-waltzing Matilda with me.'

Up rode the squatter mounted on his thorough-
bred, down came troopers – one, two, three.
'Whose is that jumbuck
you've got in the tuckerbag?'

'You'll come a-waltzing Matilda with me.'
Up jumped the swagman and sprang into the
billabong.

'You'll never catch me alive,' said he.
And his ghost may be heard
as you pass by that billabong,
'You'll come a-waltzing Matilda with me.'

'Click, click, click!'

The much-loved Australian folk song 'Click Go the Shears', the author of whom is unknown, conjures images of a typical shearing shed of the 1800s. It is in sung to the tune of an American song, 'Ring the Bell Watchman', composed in 1865 by self-taught musician, abolitionist and small-time inventor Henry Clay Work.

Out on the board the old shearer stands
grasping his shears in his thin bony hands,
fixed is his gaze on a bare-bellied yoe,
glory if he gets her, won't he make the ringer go.

Click go the shears boys, click, click, click!
Wide is his blow and his hands move quick.
The ringer looks around and is beaten by a blow,
and curses the old snagger with the bare-bellied yoe.

· ·

Did you know?

New Zealanders lead the world in Google searches for the word "sheep". They're also number two for looking up "porn". Australia is second and fourth respectively on the global list for seeking the same words on the internet search engine. By coincidence, Aussie and Kiwi men are numbers one and two at typing "girlfriend" in the search field...

· ·

A prolific writer, Banjo Paterson also penned what might be considered one of the most evocative pieces of prose of this era, The Merino Sheep:

'The Merino Sheep'
by A. B. Banjo Paterson

People have got the impression that the merino is a gentle, bleating animal that gets its living without trouble to anybody, and comes up every year to be shorn with a pleased smile upon its amiable face. It is my purpose here to exhibit the merino sheep in its true light.

First let us give him his due. No one can accuse him of being a ferocious animal. No one could ever say that a sheep attacked him without provocation; although there is an old bush story of a man who was discovered in the act of killing a neighbour's wether.

'Hello!' said the neighbour, 'What's this? Killing my sheep! What have you got to say for yourself?'

'Yes,' said the man, with an air of virtuous indignation. 'I am killing your sheep. I'll kill any man's sheep that bites me!'

But as a rule the Merino refrains from using his teeth on people. He goes to work in another way.

The truth is that he is a dangerous monomaniac, and his one idea is to ruin the man who owns him. With this object in view he will display a talent for getting into trouble and a genius for dying that are almost incredible.

If a mob of sheep see a bush fire closing round them, do they run away out of danger? Not at all, they rush round and round in a ring til the fire burns them up. If they are in a riverbed, with a howling flood coming down, they will stubbornly refuse to cross three inches of water to save themselves. Dogs may bark and men may shriek, but the sheep won't move. They will wait there til the flood comes and drowns them all, and then their corpses go down the river on their backs with their feet in the air.

A mob will crawl along a road slowly enough to exasperate a snail, but let a lamb get away in a bit of rough country, and a racehorse can't head him back again. If sheep are put into a big paddock with water in three corners of it, they will resolutely crowd into the fourth, and die of thirst.

When being counted out at a gate, if a scrap of bark be left on the ground in the gateway, they will refuse to step over it until dogs and men have sweated and toiled and sworn and 'heeled 'em up', and 'spoke to 'em', and fairly jammed them at it. At last one will gather courage, rush at the fancied obstacle, spring over it about six feet in the air, and dart away. The next does exactly the same, but jumps a bit higher. Then comes a rush of them following one another in wild bounds like antelopes, until one overjumps himself and alights on his head. This frightens those still in the yard, and they stop running out.

Then the dogging and shrieking and hustling and tearing have to be gone through all over again. (This on a red-hot day, mind you, with clouds of blinding dust about, the yolk of wool irritating your eyes, and, perhaps, three or four thousand sheep to put through). The delay throws out the man who is counting, and he forgets whether he left off at 45 or 95. The dogs, meanwhile, have taken the first chance to slip over the fence and hide in the shade somewhere, and then there are loud whistlings and oaths, and calls for Rover and Bluey. At last a dirt-begrimed man jumps over the fence, unearths Bluey, and hauls him back by the ear. Bluey sets to work barking and heeling 'em up again, and pretends that he thoroughly enjoys it; but all the while he is looking out for another chance to 'clear'. And THIS time he won't be discovered in a hurry.

There is a well-authenticated story of a shipload of sheep that was lost because an old ram jumped overboard, and all the rest followed him. No doubt they did, and were proud to do it. A sheep won't go through an open gate on his own responsibility, but he would gladly and proudly 'follow the leader' through the red-hot portals of Hades: and it makes no difference whether the lead goes voluntarily, or is hauled struggling and kicking and fighting every inch of the way.

For pure, sodden stupidity there is no animal like the Merino. A lamb will follow a bullock-dray, drawn by sixteen bullocks and driven by a profane person with a whip, under the impression that the aggregate monstrosity is his mother. A ewe never knows her own lamb by sight, and apparently has no sense of colour. She can recognize its voice half a mile off among a thousand other voices apparently exactly similar; but when she gets within five yards of it she starts to smell all the other lambs within reach, including the black ones -- though her own may be white.

> **A ewe never knows her own lamb by sight, and apparently has no sense of colour.**

The fiendish resemblance which one sheep bears to another is a great advantage to them in their struggles with their owners. It makes it more difficult to draft them out of a strange flock, and much harder to tell when any are missing.

Concerning this resemblance between sheep, there is a story told of a fat old Murrumbidgee squatter who gave a big price for a famous ram called Sir Oliver. He took a friend out one day to inspect Sir Oliver, and overhauled that animal with a most impressive air of sheep-wisdom.

'Look here,' he said, 'at the fineness of the wool. See the serrations in each thread of it. See the density of it. Look at the way his legs and belly are clothed — he's wool all over, that sheep. Grand animal, grand animal!'

Then they went and had a drink, and the old squatter said, 'Now, I'll show you the difference between a champion ram and a second-rater.' So he caught a ram and pointed out his defects. 'See here — not half the serrations that other sheep had. No density of fleece to speak of. Bare-bellied as a pig, compared with Sir Oliver. Not that this isn't a fair sheep, but he'd be dear at one-tenth Sir Oliver's price. By the way, Johnson' (to his overseer), 'what ram is this?'

The Merino relies on passive resistance for his success.

'That, sir,' replied the astounded functionary – 'that is Sir Oliver, sir!'

There is another kind of sheep in Australia, as great a curse in his own way as the Merino — namely, the cross-bred, or half-Merino-half-Leicester animal. The crossbred will get through, under, or over any fence you like to put in front of him. He is never satisfied with his owner's run, but always thinks other people's runs must be better, so he sets off to explore. He will strike a course, say, southeast, and so long as the fit takes him he will keep going southeast through all obstacles — rivers, fences, growing crops, anything. The Merino relies on passive resistance for his success; the crossbred carries the war into the enemy's camp, and becomes a living curse to his owner day and night.

Once there was a man who was induced in a weak moment to buy twenty crossbred rams. From that hour the hand of

Fate was upon him. They got into all the paddocks they shouldn't have been in. They scattered themselves over the run promiscuously. They visited the cultivation paddock and the vegetable garden at their own sweet will. And then they took to roving. In a body they visited the neighbouring stations, and played havoc with the sheep all over the district.

The wretched owner was constantly getting fiery letters from his neighbours: 'Your blanky rams are here. Come and take them away at once,' and he would have to go nine or ten miles to drive them home. Any man who has tried to drive rams on a hot day knows what purgatory is. He was threatened every week with actions for trespass.

He tried shutting them up in the sheepyard. They got out and went back to the garden. Then he gaoled them in the calf-pen. Out again and into a growing crop. Then he set a boy to watch them; but the boy went to sleep, and they were four miles away across country before he got on to their tracks.

At length, when they happened accidentally to be at home on their owner's run, there came a big flood. His sheep, mostly Merinos, had plenty of time to get on to high ground and save their lives; but, of course, they didn't, and were almost all drowned. The owner sat on a rise above the waste of waters and watched the dead animals go by. He was a ruined man. But he said, 'Thank God, those crossbred rams are drowned, anyhow." Just as he spoke there was a splashing in the water, and the twenty rams solemnly swam ashore and ranged themselves in front of him. They were the only survivors of his twenty thousand sheep. He broke down, and was taken to an asylum for insane paupers. The crossbreds had fulfilled their destiny.

The crossbred drives his owner out of his mind, but the Merino ruins his man with greater celerity. Nothing on earth will kill crossbreds; nothing will keep Merinos alive. If they are put on dry salt-bush country they die of drought. If they are put on damp, well-watered country they die of worms, fluke, and footrot. They die in the wet seasons and they die in the dry ones.

The hard, resentful look on the faces of all bushmen comes from a long course of dealing with Merino sheep. The Merino dominates the bush, and gives to Australian literature its melancholy tinge, its despairing pathos. The poems about dying boundary riders, and lonely graves under mournful she oaks, are the direct outcome of the poet's too close association with that soul-destroying animal. A man who could write anything cheerful after a day in the drafting-yards would be a freak of nature.

A wolf in sheep's clothing

Aesop, a slave in Greece during the sixth century BC, recorded many fables, one of the more well known recalling the tale of the wolf in sheep's clothing. The story goes that there was once a wolf who was rather intent – as wolves tend to be – on gaining access to a flock of sheep and a lamb-chop dinner. His way was barred, however, by the very vigilant shepherd and his sheep dogs. One day, the wolf fortuitously came across the discarded skin of a dead sheep. Thinking quickly – as wolves tend to do – the wolf slipped the skin over his own coat and, pretending to be a sheep, made his way easily through the flock, fooling both the shepherd and his sheep. The wolf's day could not have been better.

That night, the shepherd followed his usual routine and locked up his flock in the pen, including the wolf. While the wolf thought this was akin to being in heaven, he was in for a nasty shock. He wasn't the only one who enjoyed a lamb-chop dinner: later that night the shepherd returned to the pen to kill his fattest sheep for his own evening meal. And the sheep he chose? The wolf. So the wolf's day, it would seem, could not have been worse.

Sheep in the spotlight

Media daaarlings

In March 2006 *The Times* reported the Irish Republic's intention to undertake a spring clean of some of its more outdated laws, including one which denied citizens the right to add sheep dung to coffee. More curious, however, than this odd law's existence is the reason for the necessity of such a law banning people from plonking poo in your morning cuppa.

As they say, any publicity is good publicity. But is the story of an ovine suicide plunge really going to further the cause for an animal already blighted with a reputation for seeming a little dimwitted? In July 2005 international media reported on a rather bizarre event. Near the town of Gevas, in eastern Turkey, shepherds were enjoying their breakfast while their flock, which consisted of sheep belonging to 26 local families, grazed nearby. Soon the shepherds noticed one sheep jumping to its death over a cliff. Then another, and another. In all, around 1500 sheep followed one another over the cliff to what would seem certain death.

Around 450 of the animals died – the bodies of the first few hundred piled up at the bottom of the 15-metre (49 ft) drop, forming a soft cushion for those that followed and consequently saving the flock from an even bigger disaster. But although the loss could have been greater, the financial impact on the families involved was still huge, equalling around AU$142,000 (£57,000). And while the shepherds remain uncertain as to why the sheep jumped, some believe they may have been trying to leap to the other side of the small ravine where they eventually met their end.

New Zealand info website suff.co.nz reports that in July 2005 that country's Advertising Standards Complaints Board received a complaint from people concerned over the nature of a television advertisement for a Toyota ute. In the ad, two animated bulls apparently steal a ute in the hopes of impressing two cows, and while driving refer to a ram they pass as a 'sheep shagger'. It is to this phrase that the complainants took most offence. Also of concern was a scene where the ute drives through a flock of sheep and over an embankment, prompting claims that the ad displayed dangerous driving and raised concerns over animal welfare.

The Complaints Board ruled that the advertisement was not based in reality (as perhaps the complainants might not have been) and was obviously an animated fantasy, and the matter was dropped.

September 2005 saw US media reporting on a college prank involving a cross-dressing sheep. The sheep was stolen from an organization known as the Natural Resources Trust, dressed in a bra, painted black and placed in a dormitory at Stonehill College in Easton, near Boston. It was suspected by many that the college's rugby team were behind the prank.

The incident prompted the school's rugby team officials to punish their entire team by indefinitely placing them on the sidelines while investigations into finding the culprit continued. According to staff at the Natural Resources Trust, the sheep was clearly traumatised by the treatment meted out to it and never seemed to recover. It died just two months after the ordeal.

. . . .

Did you know?

In 2002 a British-based writer was given a local government grant to use sheep to create random poems.

Each of the animals had a word from a poem written on their backs. As they wandered about the words took a new poetic form each time they came to rest.

Writer Valerie Laws said: 'I like the idea of using living sheep to create a living poem, and creating new work as they move around. Randomness and uncertainty is at the centre of how the universe is put together, and is quite difficult for us as humans who rely on order. So I decided to explore randomness and some of the principles of quantum mechanics, through poetry, using the medium of sheep.'

One of the poems created by the sheep reads:

Warm drift, graze gentle, White below the sky, Soft sheep, mirrors, Snow clouds.

. . . .

Like his more famous namesake, Shrek the sheep shunned the company of others and chose instead the life of a hermit. The 10-year-old Merino ram, a resident of Bendigo hill station on New Zealand's South Island, managed to elude the shearing shed for six years, mostly, it's believed, by hiding out in a cave. Once finally caught, in April 2004, he was publicly parted from his fleece live on national television. The fleece weighed in at almost 27 kg (60 lb), estimated to be enough to make suits for 20 large men and was auctioned off to raise money for children's medical charities.

Sheep hit news headlines in a big way in August 2003 when over 57,000 of the innocent beasts became embroiled in a political and humanitarian fiasco at sea. The sheep were bound for Saudi Arabia aboard the MV *Cormo Express*, when while at sea the Saudi buyers cancelled their order, claiming to have found cases of scabby mouth disease amongst the live cargo. So began a saga that was to last for 11 weeks, while the sheep remained helplessly stranded in the Persian Gulf.

After leaving the Saudi port where the sheep had been inspected and rejected, the ship lurched from port to port, taking on fodder, while ideas as to what to do with the sheep were debated in parliament. More than 50 countries were approached by the Australian government and offered the sheep – no one took them. There was talk of the sheep being returned to Australia, and a team of government officials eventually flew to the Middle East in an attempt to secure another buyer. Animal welfare groups became involved in the argument, some urging the federal government to show compassion and have the sheep slaughtered while still at sea. Rock legend Chrissie Hynde, of UK band The Pretenders, made her way to Sydney to protest what many considered Australia's inhumane practice of live sheep exports. The saga finally ended in late October when the sheep were accepted by the Eritrean government, along with a gift of AU$1 million (£412,000) and 3000 tonnes of sheep fodder.

In the show ring

With an average annual attendance of 150,000, Scotland's Royal Highland Show is one of the UK's premier agricultural shows. Every year over 5000 animals are displayed at the show, including around 1400 proud sheep. And while the sheep shearing competition might have showmanship on its side to whip the crowd into a frenzy, the hottest competition happens in the interbreed contests. Around 20 breeds are exhibited and battle for the title of top sheep, including Beltex, Suffolk, Cheviot, Swaledale, Blackface, Texel and Shetland breeds.

Head to the Bruce Pavilion at the massive Sydney Royal Easter Show to find ovine competitors vying for various titles at Australia's largest agricultural show. Providing a much-anticipated annual Easter outing for Sydney families since the early 1820s, the Easter Show's sheep judging sections are highly competitive and the trophies and ribbons much coveted.

While the show has been running since 1823, the first animals weren't displayed until the following year, when 43 pens of sheep strutted their stuff alongside cattle, stallions, ponies, dogs and poultry. Interestingly, in the show's first year male servants were judged, with one William Bull receiving a cash prize for faithful service and good conduct. (Women servants didn't make the grade until 1825, the same year that medals replaced cash prizes, as cash was believed to encourage 'intoxication and other excesses' – presumably amongst the servants, not the livestock.)

Along with the carnival rides, the sugar-laden showbags, the fireworks and around-the-clock entertainment, today's event, which draws annual crowds of around 1 million, still showcases the best of rural life: there are woodchopping competitions; crafts displays; the finals for the Royal Agricultural Society/Stock and Station Agents Association Young Auctioneers Competition; horticulture displays; and the perennially popular district exhibits, where over 10,000 pieces of fruit and veg are meticulously arranged into complex, themed scenes by volunteers from five competing regional areas. For true blue agricultural fans, however, the livestock competitions remain centre stage at the Royal Easter Show. Cattle, pigs, goats and alpacas, horses, domestic animals and even bees are offered up by their breeders and owners for judging, along with our friends the sheep.

Top awards for sheep at the Royal Easter Show include:
- Grand Champion Border Leicester Ram
- Champion August Shorn Merino Ewe
- Best Team of Six Border Leicester Sheep
- Grand Champion Strong Merino Wool Ram
- Most Successful Dorset Horn Exhibitor
- Best Headed Suffolk
- Most Successful Superfine and Fine Wool Merino Exhibitor.

So what are the criteria by which a sheep's fleece is judged? Australia's Hamilton Sheepvention, where the ewe weaner contest hots up every year, stipulates the following amongst its guidelines for those wanting to hoof it on up to the winner's podium:
- Only those stud breeders selling less than 25 rams per year may enter, and the sheep they enter must be owned and bred by them.

- Sheep entrants must have been kept under their usual farm conditions for at least two weeks prior to the competition (i.e. not pampered in five-star accommodation, with access to a day spa and a fridge full of fine food).
- Entrants must not have more than two incisor teeth.
- The entrant's bloodline must be outlined on their entry form.

If you'd like to see wool make it from the sheep's back to yours in record time, the Pennsylvania Farm Show is the place to go. Here, the Sheep to Shawl contest is held according to strict rules. Small teams of fast-working experts take the fleece from the sheep's back and make it into a shawl, with the first team to complete their shawl being hailed the winner.

Teams consist of one shearer and five other members: the carder, three spinners and one weaver. Once the gun goes off, it's all hands on deck. The shearer removes the fleece from the team's chosen sheep, which must be neither dyed nor tranquillized, but can be washed up to two weeks prior to the competition. Then the carder – who may also be the shearer – combs out the fleece, ready to hand over to the spinners. The yarn is finally woven into a shawl that must conform to set requirements: it must be a minimum of 55.8 centimetres (22 in) wide and 198 centimetres (78 in) long, and must have a minimum of 12.7 centimetres (5 in) of fringe on each end.

A panel of four judges award teams points for shearing, spinning, weaving and design, as well as for something mysteriously labelled 'team identification'. The fastest team for 2006 was Red Rose Treadling Toes, while the perhaps portentously named Loose Threads team came in at the bottom of the ladder.

Preparing a lamb for show can be a time-consuming business. In order for the lamb's muscle definition to stand out clearly the lamb needs to be sheared as close to the skin as possible. Before this can happen the animal needs to be washed to make the wool easy to clip and dry — often using industrial livestock blowers. Shearing can then proceed, often using a variety of clippers.

The lamb should then be wrapped to ensure it stays clean This can often involve placing socks on each leg. Just before entering the ring any dirty marks are removed and the inside of the lamb's ears cleaned with a wet cloth.

A dog's life

What would a flock of sheep be without a little direction? And who better to provide that direction than the sheep farmer's best friend, the working dog. Sheep dogs, consisting of breeds such as the border collie and the kelpie, spend their working lives in the paddock, corralling and cajoling sheep with their endless energy and proving themselves to be invaluable employees. Great pride is taken in their considerable skills, which are showcased in various annual sheep dog trials.

Rules for sheep dog trials vary, but a basic outline of some of the rules set down by the West Australian Working Sheep Dog Association offers an introduction to this very competitive and complex sport. Dogs begin their time trial with a total of 100 points, from which points are deducted for errors on the part of either the dog or the owner. Each dog works three sheep in an arena. The sheep are driven by the dog through a maze of obstacles, over bridges and finally into a pen. The dog must work unassisted by their owner – however the dog may be directed by a series of whistles and signals. If any sheep escape while being worked by the dog, the trial may be terminated by the judges.

Did you know?

Sniffer dogs have been trained to tell which sheep have parasitic worms. By sniffing the poo of suspect sheep the dog can indicate if problem parasites are present. Parasites can interfere with a sheep's wool and meat growth, and make the animal weak and susceptible to other diseases.

Did you know?

Sheep used in sheep dog trials are usually classed according to temperament

Light sheep

These are sheep that are easy for a dog to move. This often makes the trial harder as they run away with little or no provocation.

Heavy sheep

These are stubborn sheep that are difficult for a dog to move. They will sometimes even attack another dog.

Dogged sheep

After being used repeatedly for training sheep dogs, sheep can become 'dogged'. Dogged sheep will rush to the handler as soon as the dog is sent off to fetch them. They will crowd around the handler's legs and can become extremely difficult to work with.

Hobby sheep

If you've got some available space, and are looking for a great
alternative to a family dog or cat, you might like to consider
keeping a sheep or two as a pet. With their renowned gentle
nature, many who already keep them say they are wonderful
companion animals and offer a great way for children
to learn the responsibility and joy in caring for another
creature. Following are a few basic tips to get you started.

- While almost any breed of sheep can be kept as a pet, you
  should choose either a female lamb, a ewe, or a polled
  wether (an older, castrated male without horns).
- As with all animals, whether pets or livestock, a clean
  supply of water is essential.
- If you've decided on a pet lamb, not only will you need to
  keep in mind your cuddly new friend will one day be a fully
  grown sheep, but there are some special responsibilities
  involved. Newborn lambs need to be fed with a bottle, and
  there are commercial milk powders available specifically for

lambs. You'll also need to make sure your lamb is kept clean and free of any blow flies.

- Sheep need quality roughage to eat, in the form of grass and clover, and you can also feed them hay. They mostly prefer young, more tender plants, and will graze at dusk and dawn for a total of around seven hours per day. If the hay or grass available to you is poor, you can supplement this with some special sheep feed mix. Providing a salt block for added minerals is also a good idea.
- Your sheep or lamb will need enough room to exercise sufficiently. If using a pen, set up some logs or poles and encourage your sheep to jump over them.
- You should provide some kind of shelter from the sun, rain and cold. If you plan to keep a lamb, you'll need to provide a warm area for them during the colder months, as lambs lose body heat more easily than grown sheep. You might also like to provide some straw bedding, particularly if you live in a cool climate.
- Your pet will need to be sheared annually, and it's recommended that a professional do this for you. If you'd rather not bother, however, consider getting a Wiltshire Poll, as the breed naturally sheds much of its wool each year, thus reducing the need for shearing.
- Like most other pets, sheep need to be wormed and vaccinated – check with your vet for more details on the kinds and frequency of vaccination required. Also, hooves need to be kept trim and occasionally checked for 'rot'.

If you're serious about keeping a sheep as a pet, do your homework before you bring home the newest addition to your family. There are numerous books on sheep husbandry for the hobby farmer which will give you all the detailed information you need.

Folk Sheep, festive sheep

You conclude that my master is a shepherd then, and I a sheep? I do. Well then, my horns are his horns, whether I wake or sleep. A silly answer, and fitting well a sheep.

– *Two Gentlemen of Verona,*
William Shaspeare, act I scene I

Beliefs, superstitions, folklore

In Romanian folklore, clouds are divided into those closer to the earth and those further away. The 'dark' or 'buffalo' rain clouds are the closer ones, and are believed to be part of the revenge of the wizard Solomonari on the earth. The highest clouds of all are said to bring good weather and are known, among other things, as sheep clouds.

An old Manx belief holds that sheep can only be counted properly if the person doing the counting has first washed their eyes under running water.

In Iceland, it's believed that if sheep gnash their teeth during the autumn round-up, the following winter will be very hard.

Irish folklore offers a cure for an earache: simply take some wool from a black sheep and stuff it into your ear, and the earache is said to disappear.

Meeting sheep at the beginning of a journey is said to signal good luck.

In ancient Egypt, rams were associated with many gods, including Arsaphes, Khnum and Amun. It is believed rams were worshipped at temples both as a symbol of fertility and for their warlike qualities. The entrance to the temple of Amun, at Thebes, is flanked by sphinxes with the heads of rams. Some rams were even mummified and adorned in masks and jewellery.

In Turkey, to dream of a white sheep indicates snow is on the way.

The Qiang people, indigenous inhabitants of areas west of the Yellow and Yangtze Rivers in China, are sometimes referred to as the sheep people due to their sheep totems. When they put on their traditional clothing, they include head ornaments shaped like sheep's horns.

. . . .

Did you know?

In central Asia sheep are a central part of the Kazak peoples' existence and this is reflected in many of their traditions. One such tradition is a ceremony involving a boiled sheep's head. The head is presented to the oldest, or most honourable, guest. The guest returns a slice of sheep's cheek to the host as a mark of respect before serving one of the ears to the youngest guests urging them to be careful, and the sheep's palate to the young women present, who are reminded to be diligent.

Your status at the meal can be further indicated by which cut of meat you receive. Honoured guests are treated to gammon and shank. Young brides get brisket; married women, neck bones. Children are given kidney and heart, to encourage maturity. Sheep's brain is not given to children as it is believed to make them weak willed. Knuckle is never served to a young girl otherwise she may never marry.

. . . .

Yan Tan Tethera

Want to know how many births, deaths and runaways there have been in the flock? Then you have to count your sheep. Conducted at both ends of the day, the ritual of counting sheep has been a fundamental part of the shepherd's day for centuries.

From as far back as medieval times British farmers were allowed access to common grazing property, and so that this land was not overgrazed each shepherd had to keep an accurate record of how many sheep he had. Rather than using the numerical system we know today, sheep were counted using Yan Tan Tethera, a numeric jargon derived from the now-dead Cumbric language. It's believed that this repetitive, monotonous task may have been the source for the classic insomnia remedy of counting sheep.

But the fabled insomnia-beater doesn't work – or so say researchers at Oxford University. A group of 50 people were observed for an Oxford study, all of whom had trouble getting to sleep: one-third of them were asked to count sheep to try to get to sleep; another third were asked to visualize a waterfall or beach; and the final third acted as a placebo group, not being allowed to do either. Of the three groups, those who pictured the beach or waterfall fell asleep an average of 20 minutes earlier than those in the other two groups. And the ones who counted sheep? They ended up staying awake longer than they usually would.

If you don't believe the Oxford dons and still think that counting sheep will send you off blissfully to the land of nod, then you might like to invest in 'Counting Sheep: Fall asleep the natural way', a DVD that will apparently show you 'many small and large, cartoon and real, spotted and white sheep that ceaselessly jump past'. Sound ceaselessly dull? Perhaps you're better to simply turn off that DVD and go to bed.

Sheepish Chinese stars

In Chinese astrology, you can consider yourself a sheep if you were born in any of the following years: 1907, 1919, 1931, 1943, 1955, 1967, 1979, 1991, 2003. Also known as the goat sign, Chinese sheep are a little different from their Western astrological counterparts; they're said to be charming and amiable, and constantly looking for the best in other people. They are also reputedly wise, elegant and quite accomplished in the arts. Sound too perfect? Don't worry, they're also clumsy and timid, and not particularly practical or forceful, and they need the influence of a firm hand in their life. Famous Chinese sheep include Marcel Proust, Phillip Seymour Hoffman, Joni Mitchell, Bill Gates, Michelangelo and Charles Dickens.

. . . .

Did you know?

The famous English meteorological proverb stating, 'When March comes in like a lion it goes out like a lamb' is based on scientific fact. The temperature difference between the equator and the poles is at its maximum in March and this often creates ideal conditions for rain and storms which seldom last very long, fading out before the month's end.

. . . .

Proverbs, sayings and expressions

There are numerous Chinese expressions referring to sheep. To describe someone or something as like 'a camel amidst a flock of sheep' implies that they stand out in an awkward way. And if you mend the pen only after the sheep are all gone, you're not doing much better than having closed the gate once the horse has bolted. This proverb, however, has a further meaning: some believe it's better to mend the pen once the sheep have gone rather than not at all – a lesson is then learned from having lost the sheep. It's also said by the Chinese that a sheep has no choice when in the jaws of a wolf.

An army of sheep led by a lion would defeat an army of lions led by a sheep. (Arab)

Without a shepherd, sheep are not a flock. (Russian)

There was never a scabby sheep in a flock that didn't like to have comrades. (Irish)

A leap year is never a good sheep year. (English)

You might as well be hanged for a sheep as for a lamb. (English)

Did you know?
The word 'sheep' is pronounced in a similar way to 'auspicious' in Chinese.

Dressed for a party

Each October in the Idaho towns of Ketchum and Hailey, the sheep are let loose to wreak havoc. The Trailing of the Sheep Festival is optimistically billed on its official website as 'America's version of the Running of the Bulls', and recalls a slower, more rural life. For over 100 years, shepherds have moved their flocks from higher summer pastures down through the local towns to graze for the winter, and the festival, begun in 1997, honours this part of the region's history. Over 2000 sheep slowly meander through the towns' main thoroughfares, past shops and cafes, closely followed by shepherds, sheep wagons and visitors.

The festival celebrates all things ovine. Spread over three days, events also include lamb cooking classes, workshops on spinning and weaving, a fair showcasing shepherding skills and regional culture, and the chance to sample specially prepared lamb recipes at local restaurants.

The Eid al-Adha festival is held in the last month of the Islamic calendar, when Muslims from around the world make a pilgrimage to Mecca. One of the most anticipated celebrations of the year, the festival honours the prophet Abraham's willingness to sacrifice his son, as demanded by Allah. Once Abraham had shown his devotion through this act, Allah replaced Ishmael with a sheep to be sacrificed.

Friends and family visit each other, children receive gifts, and food is shared, with lamb and mutton forming the most important dishes on the menu to honour Abraham's devotion. Families often save for many months to afford a sheep or goat to be slaughtered for the festival. In keeping with the festival's spirit of giving, one-third of the meat is kept by the family, one-third is given to relatives, and the final third is given to the community's poor.

Folks have been gathering in celebratory style in the picturesque English village of Findon, West Sussex, since as far back as 1261. What began as a three-day fair had been distilled by 1650 into a one-day event. The local sheep fair began sometime in the 1780s and incorporated lamb sales, with the first auctions being held in 1896. And for such a small community, the size of these auctions certainly grew: by the 1920s the number of sheep auctioned at Findon had swelled to around 10,000, and the fair attracted people from miles away.

While what is now called the Findon Sheep Fair and Village Festival has continuously hosted a sheep auction for over 100 years, it seems that such an auction is becoming increasingly financially untenable, and so 2006 sees the end of this part of the fair's history. The festival, however, continues to thrive in other areas, offering visitors a traditional fairground, family dances, falconry exhibitions and Punch and Judy shows. And while the sheep are no longer being auctioned off, their presence is still an important part of the festival, with a talk on the history of sheep and shearing, and sheep dog demonstrations rounding out the entertainment.

If you're looking to hold your own ovine festival but don't know where to start, give the guys at the Sheep Show a call. Based in the United Kingdom, New Zealanders Richard

Savory and Stuart Barnes take their travelling show around the country, setting up their self-contained semi-trailer stage and putting on a 30-minute show which introduces nine breeds of sheep to the audience and educates them about the sheep and their wool. Shearing demonstrations are also given.

. . . .

Did you know?
The Yi people in Guizhou, China, use a sheep's stomach and a gourd to make their traditional wine bottles.

. . . .

Quotable quotes

'Every man can tell how many goats or sheep he possesses, but not how many friends.'
– Cicero, Roman politician, 106–43 BC

It's better to be a lion for a day
than a sheep all your life.
— Sister Elizabeth Kenny,
Australian nurse, 1880–1952

In order to be a member of an immaculate flock
of sheep, one must above all be a sheep oneself.
— Albert Einstein, German physicist, 1879–1955

Without tradition, art is a flock
of sheep without a shepherd.
Without innovation, it is a corpse.
— Winston Churchill,
British prime minister, 1874–1965

We are discreet
sheep; we wait to
see how the drove is
going, and then go
with the drove.
— Mark Twain,
American writer, 1835–1910

Families have their fools
and their men of genius,
their black sheep and
their saints, their worldly
successes and their
worldly failures.
— Aldous Huxley, British author,
1894–1963

We herd sheep, we drive cattle, we lead people.
Lead me, follow me, or get out of my way.
— General George S. Patton, American General in
World Wars I and II, 1885–1945

If someone wants a
sheep, then that means
that he exists.
— Antoine de Saint–Exupery,
French pilot, writer and
author of 'The Little Prince', 1900–)

Every man can tell how many goats or sheep
he possesses, but not how many friends.
— Cicero, Roman politician, 106–43 BC)

To create man was a quaint and original
idea, but to add the sheep was tautology
— Mark Twain,
American writer, 1835–1910

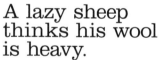

A lazy sheep
thinks his wool
is heavy.
— Turkish proverb

**If the freedom of speech is taken away then
dumb and silent we may be led, like sheep to
the slaughter.**
— George Washington, 1st US president, 1732–99

Who makes
himself a sheep
will be eaten by
the wolves.
— proverb

It never troubles the wolf how
many the sheep may be.
— Virgil, Roman poet, 70–19 BC

Better to live one
year as a tiger, than
a hundred as sheep.
— Madonna, American actress
and singer, b.1958

A society of sheep must in time beget a government of wolves.
— Bertrand de Jouvenel

A wolf eats sheep but now and then,
ten thousands are devoured by men.
— Benjamin Franklin, US inventor. 1706–90

Without a shepherd,
sheep are not a flock.
— Russian proverb

He that makes himself a sheep
shall be eaten by a wolf.
— Italian proverb

Hang out a sheep's head to sell dog's meat.
— Chinese Proverb

Without tradition, art is a flock of sheep without
a shepherd. Without innovation, it is a corpse.
— Winston Churchill,
British prime minister, 1874–1965

And when he cometh home, he calleth together his friends and neighbours, saying unto them; Rejoice with me, for I have found my sheep which was lost.

— Bible

An army of sheep led by a lion would defeat an army of lions led by a sheep.

— Arabian proverb

The sheep has no choice when in the jaws of a wolf.

— Chinese proverb

In order to be an immaculate member of a flock of sheep, one must above all be a sheep oneself.

— Albert Einstein, German physicist, 1879–1955

The shepherd always tries to persuade the sheep that their interests and his own are the same.

— Stendhal, French writer, 1783–1842

March comes in like a
lion and goes out like
a lamb.
— English proverb

A camel standing amidst a flock of
sheep looks awkward.
— Chinese proverb

Democracy must be
something more than two
wolves and a sheep voting
on what to have for dinner.
— James Bovard

It is the part of a good shepherd to shear
his flock, not to skin it.
— Latin proverb

The shepherd drives the wolf from the sheep's
throat, for which the sheep thanks the shepherd
as his liberator, while the wolf denounces him
for the same act as the destroyer of liberty.
— Abraham Lincoln
16th US president, 1809–65

Conclusion

So, after this excursion in and around a creature that has clearly been a crucial part of our lives for generations, why are sheep still so seemingly maligned? Maybe over time, as our connection with the land has lessened, our knowledge of and appreciation for them has dwindled. Or perhaps it's simply that we have a need to make fun of someone or something, so the poor defenceless sheep, as good a target as any, cops our ire.

Hopefully in reading this book you've picked up a few facts about sheep that have elevated their status a notch or two in your opinion, and with any luck you'll find sheep just a bit more interesting than you did before. You might even consider them good luck the next time you pass a flock in a field; you might quietly thank them for the warmth and versatility of the wool that makes up your favourite jumper; or you may look to the skies and see those bearers of good weather, the high sheep clouds of Romanian folklore.

Ewe be the judge.

We would love to hear your personal sheep stories for future editions of this book. Please write to sheep@murdochbooks.com

References

ABC News, 'Dog Sniffs Sheep for Worms'
http://www.abc.net.au/science/news/stories/s1617432.htm

About Turkey
about-turkey.com

Adam, D, 'Forget the tiger – put sheep urine in your tank', The Guardian, 10 June 2005.

AgResearch 2000
http://www.agresearch.co.nz/scied/search/biotech/print_gene_gmocasesheep1.htm

American Sheep Industry Association
http://www.sheepusa.org/

BBC Channel 4
http://www.channel4.com/culture/microsites/H/hirst/for.html

BBC History
http://www.bbc.co.uk/history/timelines/

BBC News, 'Nursery Rhyme Ban Scrapped'
http://news.bbc.co.uk/2/hi/uk_news/education/600470.stm

BBC News, 'How to Turn Sheep Into Art'
http://news.bbc.co.uk/2/hi/uk_news/wales/south_west/3132690.stm

BBC News, Wednesday, 4 December, 2002

Bible Topics
www.bible-topics.com
Breeds of Livestock
http://www.ansi.okstate.edu/breeds/sheep/

Burns Country
www.robertburns.org

Busack, M 'Abused sheep found in dorm in 2d similar case at Stonehill', Boston Globe, 11 September 2005.

Clark, D 2004, Big Things: Australia's amazing roadside attractions, Penguin Books, Melbourne.

Comcast News
www.comcast.net

Cotswold Sheep Society
http://www.cotswoldsheep.org/

CSIRO
www.csiro.au

Elliott, V 'How 19th century farmers used to celebrate the fat of the land', The Times, 20 March 2006.

Everything Chinese
www.everything-chinese.com

Famous Proverbs
www.1-famous-quotes.com

Findon Sheep Fair and Village Festival
www.findonsheepfair.co.uk

Guinness World Records
www.guinnessworldrecords.com

Inner Mongolia News
nmgnews.com.cn

International Wool Textile Organisation
http://www.iwto.org

Islamic Garden
http://www.islamicgarden.com/page1021.html

Kang, S 'Fashion—Ugh! Ugg backlash', Washington Examiner, 12 December 2005.

About Turkey
about-turkey.com

Manx Loaghtan Produce Company Ltd
http://www.manxloaghtan.com/

McDougall, L, 'Revealed: the proud history of haggis hurling was just a hoax', Sunday Herald, 25 January 2004.

National Gallery of Victoria
www.ngv.vic.gov.au

National Library of Australia
www.nla.gov.au

New Zealand Film Commission
www.nzfilm.co.nz

New Zealand Ministry of Agriculture and Forestry
http://www.maf.govt.nz/mafnet/rural-nz/statistics-and-forecasts/sonzaf/2003/2003-sonzaf-20.htm

North Carolina State University College of Agriculture and Life Sciences
www.cals.ncsu.edu

Oklahoma State University Department of Animal Science
http://www.ansi.okstate.edu/

Pennsylvania Farm Show
www.agriculture.state.pa.us/farmshow/site/default.asp

Ponting, K, Sheep of the World, Blandford Press, Dorset, UK, 1980.

Purdue University Department of Animal Sciences
http://ag.ansc.purdue.edu

Roach, J, 'To save sagebrush, researchers unleash the power of sheep', http://news.nationalgeographic.com/news/2005/09/0927_050927_sheep_pulse.html, 27 September 2005.

The Science Museum
www.sciencemuseum.org.uk

Seven Sisters Sheep Centre
www.sheepcentre.co.uk

Sharrock, D, 'Ireland casts off colonial legacy of crackpot laws', The Times, 13 March 2006.

The Sheep Show
www.thesheepshow.co.uk

Sheepvention, Hamilton
www.sheepvention.com

The Soay Sheep Society
http://www.soaysheepsociety.org.uk/

Statistics New Zealand
http://www.stats.govt.nz/quick-facts/industries/pastoral-agri.htm

Stuff
www.stuff.co.nz

Trailing of the Sheep Festival
www.trailingofthesheep.org

UNESCO
www.unesco.org

West Australian Working Sheep Dog Association
www.westausworkingsheepdog.com

Wikipedia
http://en.wikipedia.org

The Woolmark Corporation
www.wool.com.au

Working Sheep Dog
www.theworkingsheepdog.co.uk

World of Quotes
www.worldofquotes.com

Year of Sheep, Culture of Sheep
www.tpecc.org/newyear/sheep%20essay_n.htm

First published in 2006 by Pier 9,
an imprint of Murdoch Books Pty Limited

Murdoch Books Pty Limited Australia
Pier 8/9, 23 Hickson Road, Millers Point NSW 2000
Phone: +61 (0) 2 8220 2000 Fax: +61 (0) 2 8220 2558
www.murdochbooks.com.au

Concept and Design: Anne Marie Cummins
Main Illustrations: Danny Snell
Additional Illustrations: Anne Marie Cummins and Justin Thomas
Additional Research: Monica Berton, Louise Dodds-Ely, Monika Paratore
Production: Monika Paratore, Joanna Byrne
Original Concept: James Mills-Hicks

Chief Executive: Juliet Rogers
Publishing Director: Kay Scarlett

National Library of Australia Cataloguing-in-Publication Data
is available for this title

Printed in China by Midas Printing in 2006.

Email your sheep truth and trivia to sheep@murdochbooks.com.au.

About the author

Karen Gee is a writer and editor who lives in the Blue
Mountains, west of Sydney, in New South Wales, Australia.
She has been a keen animal lover since childhood, and her
affection and appreciation for sheep began with the ovine
characters featured in the countless children's books she
devoured when young.

Since writing this book, Karen has been eyeing off a sunny
corner of her property, beyond the vegie patch and the fruit
trees, where she is considering keeping a pet sheep or two.